# CHI-SQUARED DATA ANALYSIS AND MODEL TESTING FOR BEGINNERS

# Chi-Squared Data Analysis and Model Testing for Beginners

Carey Witkov

*Harvard University*

Keith Zengel

*Harvard University*

OXFORD

UNIVERSITY PRESS

# OXFORD

UNIVERSITY PRESS

Great Clarendon Street, Oxford, OX2 6DP,
United Kingdom

Oxford University Press is a department of the University of Oxford.
It furthers the University's objective of excellence in research, scholarship,
and education by publishing worldwide. Oxford is a registered trade mark of
Oxford University Press in the UK and in certain other countries

© Carey Witkov and Keith Zengel 2019

The moral rights of the authors have been asserted

First Edition published in 2019

Impression: 1

Published in the United States of America by Oxford University Press
198 Madison Avenue, New York, NY 10016, United States of America

British Library Cataloguing in Publication Data
Data available

Library of Congress Control Number: 2019941497

ISBN 978–0–19–884714–4 (hbk.)
ISBN 978–0–19–884715–1 (pbk.)

DOI: 10.1093/oso/9780198847144.001.0001

Printed and bound by
CPI Group (UK) Ltd, Croydon, CR0 4YY

Links to third party websites are provided by Oxford in good faith and
for information only. Oxford disclaims any responsibility for the materials
contained in any third party website referenced in this work.

# Preface

Recent ground-breaking discoveries in physics, including the discovery of the Higgs boson and gravitational waves, relied on methods that included chi-squared analysis and model testing. *Chi-squared Analysis and Model Testing for Beginners* is the first textbook devoted exclusively to this method at the undergraduate level. Long considered a "secret weapon" in the particle physics community, chi-squared analysis and model testing improves upon popular (and centuries old) curve fitting and model testing methods whenever measurement uncertainties and a model are available.

The authors teach chi-squared analysis and model testing as the core methodology in Harvard University's innovative introductory physics lab courses. This textbook is a greatly expanded version of the chi-squared instruction in these courses, providing students with an authentic scientist's experience of testing and revising models.

Over the past several years, we've streamlined the pedagogy of teaching chi-squared and model testing and have packaged it into an integrated, self-contained methodology that is accessible to undergraduates yet rigorous enough for researchers. Specifically, this book teaches students and researchers how to use chi-squared analysis to:

- obtain the best fit model parameter values;
- test if the best fit is a good fit;
- obtain uncertainties on best fit parameter values;
- use additional model testing criteria;
- test whether model revisions improve the fit.

Why learn chi-squared analysis and model testing? This book provides a detailed answer to this question, but in a nutshell:

In the "old days" it was not uncommon (though mistaken even then) for students and teachers to eyeball data points, draw a line of best fit, and form an opinion about how good the fit was. With the advent of cheap scientific calculators, Legendre's and Gauss's 200 year old method of ordinary least squares curve fitting became available to the masses. Graphical calculators and spreadsheets made curve fitting to a line (linear regression) and other functions feasible for anyone with data.

Although this newly popular form of curve fitting was a dramatic improvement over eyeballing, there remained a fundamental inadequacy. The scientific method is the testing and revising of models. Science students should be taught how to test models, not just fit points to a best fit line. Since models can't be tested without assessing measurement uncertainties and ordinary least squares and linear regression don't account for uncertainties, one wonders if maybe they're the wrong tools for the job.

The right tool for the job of doing science is chi-squared analysis, an outgrowth of Pearson's chi-square testing of discrete models. In science, the only things that are real are the measurements, and it is the job of the experimental scientist to communicate to others the results of their measurements. Chi-squared analysis emerges naturally from the principle of maximum likelihood estimation as the correct way to use multiple measurements with Gaussian uncertainties to estimate model parameter values and their uncertainties. As the central limit theorem ensures Gaussian measurement uncertainties for conditions found in a wide range of physical measurements, including those commonly encountered in introductory physics laboratory experiments, beginning physics students and students in other physical sciences should learn to use chi-squared analysis and model testing.

Chi-squared analysis is computation-intensive and would not have been practical until recently. A unique feature of this book is a set of easy-to-use, lab-tested, computer scripts for performing chi-squared analysis and model testing. MATLAB® and Python scripts are given in the Appendices. Scripts in computer languages including MATLAB, Python, Octave, and Perl can be accessed at HYPERLINK "https://urldefense.proofpoint.com/v2/url?u=http-3A__www.oup.co.uk_companion_chi-2Dsquared2020&d=DwMFAg&c=WO-RGvefibhHBZq3fL85hQ&r=T6Yii733Afo_ku3s_Qmf9XTMFGvqD_VBKcN4fEIAbL4&m=Q09HDomceY3L7G_2mQXk3T9K1Tm_232yEpwBzpfoNP4&s=DF05tbd6j37MCerXNmVwfVLR5pA7pXaEx291hsfBak4&e="www.oup.co.uk/companion/chi-squared2020.

Another unique feature of this book is that solutions to all problems are included in an appendix, making the book suitable for self-study.

We hope that after reading this book you too will use chi-squared analysis to test your models and communicate the results of your measurements!

---

# Acknowledgments

All textbook authors owe debts to many people, from teachers to supporters.

The innovative Harvard introductory physics lab course sequence, Principles of Scientific Inquiry (PSI lab), that trains undergraduate students to think like scientists using chi-squared model testing, was innovated by physics professors Amir Yacoby and Melissa Franklin more than a decade ago. Important early contributions to PSI lab were also made by instructors and instructional physics lab staff Robert Hart, Logan McCarty, Joon Pahk, and Joe Peidle. Both authors have the pleasure each semester of teaching PSI lab courses with Professors Yacoby and Bob Westervelt, Director of Harvard's Center for Nanoscale Systems. We also gratefully acknowledge the helpful feedback provided by our many students, undergraduate Classroom Assistants and graduate Teaching Fellows over the years.

We are indebted to colleagues for reviewing this book and making valuable suggestions, including Bob Westervelt. Special thanks go to Preceptor Anna Klales, Instructional Physics Lab Manager Joe Peidle, and Lecturer on Physics and Associate Director of Undergraduate Studies David Morin, who made detailed suggestions (nearly all of which were adopted) for every chapter.

The authors, while thanking many people above for their contributions and feedback, blame no one but themselves for any errors and/or omissions.

Below are some author-specific acknowledgements.

CW: I thought I understood chi-squared analysis and model testing until I met Keith Zengel. Keith's willingness to share his in-depth understanding of the technicalities of chi-squared analysis and model testing over the past three years we've taught together is greatly appreciated and gratefully acknowledged. I'd like to thank my 12-year-old son Benjamin (who decided sometime during the writing of this book that he doesn't want to be a physicist!) for not complaining (too much) over time lost together while this book was being written.

KZ: I used chi-squared analysis a lot of times as a graduate student at the ATLAS Experiment without really understanding what I was doing. Later, I had to do a lot of secretive independent research to be able to answer the good questions posed by Carey Witkov when we started teaching it together. I think we managed to learn a good amount about the subject together, mostly motivated by his curiosity, and I hope that others will benefit from Carey's vision of science education, where students learn how to speak clearly on behalf of their data.

# Contents

# 1

# Introduction

Say you discover a phenomenon that you think is interesting or important or by some miracle both. You want to understand and hopefully use your discovery to help the world, so using your knowledge of science and mathematics you derive a simple mathematical model that explains how the phenomenon works. You conclude that there is some variable "$y$" that depends on some other variable "$x$" that you can control. The simplest version of a model is that they are the same, that $y = x$. You probably aren't this lucky, and other parameters that you don't control, like "$A$" and "$B$," show up in your model. A simple example of this is the linear model $y = Ax + B$. You don't control $A$ or $B$, but if you had some way of estimating their values, you could predict a $y$ for any $x$. Of course, in real life, your model may have more variables or parameters, and you might give them different names, but for now let's use the example of $y = Ax + B$.

What we want is a way to estimate the parameter values ($A$ and $B$) in your model. In addition to that, we'd like to know the uncertainties of those estimates. It takes a little extra work to calculate the uncertainties, but you need them to give your prediction teeth—imagine how difficult it would be to meaningfully interpret a prediction with 100% uncertainties! But even that isn't enough. We could make parameter estimates for any model, but that alone wouldn't tell us whether we had a *good* model. We'd also like to know that the model in which the parameters appear can pass some kind of experimental "goodness of fit" test. A "go/no go" or "accept/reject" criterion would be nice. Of course, there's no test you can perform that will provide a believable binary result of "congratulations, you got it right!" or "wrong model, but keep trying." We can hope for relative statements of "better" or "worse," some sort of continuous (not binary) metric that allows comparison between alternatives, but not a statement of absolute truth.

Summarizing so far, we've specified four requirements for a method of analyzing data collected on a system for which a model is available:

1. Parameter estimation (often called *curve fitting*).
2. Parameter uncertainty estimation.
3. Model rejection criterion.
4. Model testing method with continuous values.

*Chi-Squared Data Analysis and Model Testing for Beginners*. Carey Witkov and Keith Zengel.
© Carey Witkov and Keith Zengel 2019. Published in 2019 by Oxford University Press.
DOI: 10.1093/oso/9780198847144.001.0001

Since the general problem we presented (parameter estimation and model testing) is a universal one and since the specific requirements we came up with to solve the problem are fairly straightforward, you might expect there to be a generally accepted solution to the problem that is used by everyone. The good news is that there is a generally accepted method to solve the problem. The bad news is that it is not used by everyone and the people who do use it rarely explain how to use it (intelligibly, anyway).

That's why we wrote this book.

This generally accepted solution to the system parameter estimation and model testing problem is used by physicists to analyze results of some of the most precise and significant experiments of our time, including those that led to the discoveries of the Higgs boson and gravitational waves. We teach the method in our freshman/sophomore introductory physics laboratory course at Harvard University, called *Principles of Scientific Inquiry*. Using the software available at this book's website to perform the necessary calculations (the method is easy to understand but computation intensive), the method could be taught in high schools, used by more than the few researchers who use it now, and used in many more fields.

The method is called *chi-squared analysis* ($\chi^2$ analysis, pronounced "kai-squared") and will be derived in Chapter 3 using the maximum likelihood method. If in high school or college you tested a hypothesis by checking whether the difference between expected and observed frequencies was significant in a fruit fly or bean experiment, you likely performed the count data form of chi-squared analysis introduced by statistician Karl Pearson. Karl Pearson's chi-squared test was included in an American Association for the Advancement of Science list of the top twenty discoveries of the twentieth century.[1] The version of chi-squared analysis that you will learn from this book works with continuous data, the type of the data you get from reading a meter, and provides best fit model parameter values and model testing. Parameter estimation and model testing, all from one consistent, easy to apply method!

But first, let's start with a practical example, looking at real data drawn from an introductory physics lab experiment relating the length and period of a pendulum. The period of a simple pendulum depends on its length and is often given by $T = 2\pi\sqrt{\frac{L}{g}}$. This pendulum model isn't linear, but it can be cast in the linear form $y = Ax$ as follows: $L = \frac{g}{4\pi^2}T^2$, where $X = T^2$, $Y = L$, and $A = \frac{g}{4\pi^2}$. However, something seems to be wrong here. In this experiment we change the length ($L$) and observe the period ($T$), yet $L$ in the model is the dependent variable $y$! There's a reason for this. **One of the simplifications we will be using throughout most of this book is the assumption that all uncertainties are associated with measurements of the dependent variable $y$.** You might think that measurements of pendulum length should have even less uncertainty

---

[1] Hacking, I. (1984). *Science* **84**, 69–70.

than pendulum period, but we assure you that from experience teaching many, many, physics lab sections, it doesn't! The accuracy of photogate sensors or video analysis, coupled with the large number of period measurements that can be averaged, makes the uncertainty in period data fairly small, while meter stick measurements, especially when long pendulums (greater than a meter) are involved, often exhibit surprisingly large uncertainties.

Figure 1 displays a plot of data collected over a wide range of pendulum lengths and periods. There are two important facts to note about the plot. First, each data point represents a mean value, not an individual trial value, so each data point is the result of repeated measurements. Second, the plot line is not the best fit line. Instead, the plot line is the model line calculated using known parameter values. This is a model where we have the luxury of knowing parameter values precisely ($\pi$, 2, and the acceleration due to gravity, $g$) before performing any experiments on our system.

Looking at the data points and model line in figure 1, what are your thoughts about how well the data fits the model line? At first glance the fit looks pretty good, right? A naive approach would be to assume from the close visual fit that the model must be good and then proceed to obtain a best fit line using linear regression (least squares with a linear model).

Least squares goes back more than two centuries to Legendre and Gauss and involves minimizing the sum of the squared differences between each data point and the model,

$$\sum_i (y_i - y_{\text{model}})^2,$$

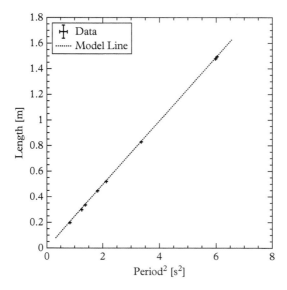

**Figure 1** *Pendulum data and model line.*

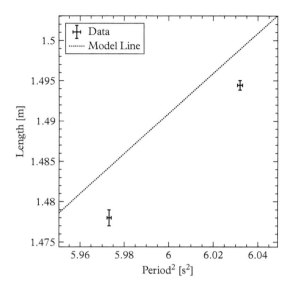

**Figure 2** *Pendulum data with error bars after zooming.*

to arrive at best estimates for parameters $A$ and $B$ in the model of a straight line $y = Ax + B$.

By zooming in on a few of the data points, as seen in Fig. 2, we see that what initially looked like a good fit is not. Each error bar corresponds to plus or minus one standard error ("one sigma" or $1\sigma$).[2] It is now clear that the discrepancy between data and model is big enough to drive a (several $\sigma$) truck through! Note that some data points are many sigmas ($\sigma$'s) from the model. A discrepancy of only two sigma is sometimes considered sufficient to reject a model. In this case the model was entirely wrong, but we couldn't tell simply by "eyeballing" it.

It might seem like any *curve fitting* technique could constitute a form of model testing, but, as shown in this example, failure to include measurement uncertainties can prevent us from seeing that the fit to the data is poor. For example, linear regression, which is built into even the least expensive scientific calculators and is one of the most common uses for spreadsheets in business and science, does not include measurement uncertainties. Therefore, ordinary least squares is not an acceptable method for parameter estimation because it treats all measurements the same, even if their uncertainties vary greatly. Tacking other methods onto ordinary least squares (like correlation coefficients or the F-test) in an effort to add some form of model testing is a fragmented approach that doesn't address the key issues: that parameter estimation and model testing require measurement uncertainties and that parameter estimation and model testing cannot be separated. Parameter

---

[2] The standard error is the standard deviation of the mean. Standard deviation and standard error are widely used measures of uncertainty that are formally introduced in the Chapter 2.

estimation and model testing is a "chicken or egg" problem: how can we know if a model is good without first estimating its parameters, yet how can we estimate meaningful model parameters without knowing if the model is good?

In this book we present one consistent methodology, chi-squared analysis, built on the *terra firma* of probability theory, that combines parameter estimation and model testing by including measurement uncertainties and answers five important questions about the results of an experiment on a system for which a model is available:

1. What are the best fit parameter values?
2. Is the best fit a good fit?
3. What are the uncertainties on the best fit parameters?
4. Even if the fit is good, should the model still be rejected?
5. Is the revised model an improvement?

# 2
# Statistical Toolkit

## 2.1 Averaging

Chi-squared analysis requires familiarity with basic statistical concepts like the mean, standard deviation, and standard error. Even if you are familiar with these terms it would still be a good idea not to skip this chapter as we present them in a way that is most useful for chi-squared analysis and discuss some novel aspects.

Let's start with a simple reaction time experiment: you hold a vertical meter stick out in front of you while a friend (or a lab partner if you don't have any friends) holds their open hand near the zero mark at the bottom, then you drop it and they catch it as quickly as they can. You and your friend (or whoever) can estimate their reaction time from this experiment using the kinematic relation between displacement and time for constant acceleration with zero initial position and zero initial velocity,

$$d = \frac{1}{2}gt^2, \tag{1}$$

which can be rearranged to find[1]

$$t = \sqrt{\frac{2d}{g}}. \tag{2}$$

If you perform the experiment twenty times, discard the first trial to avoid measuring "practice effects,"[2] starting with your hand half open, you might obtain results like those in Table 1, for five trials.

---

[1] We're going to go ahead and throw away the negative reaction times that result from the square root here, but it's probably worth noting that at least once in physics history a Nobel Prize was awarded to someone who refused to disregard apparently non-physical negative values. In 1929, Paul Dirac interpreted positive and negative energies resulting from a square root as corresponding to particles and anti-particles. He was awarded the Nobel Prize in 1933, and Carl Anderson, who discovered the first anti-particle in 1932, was awarded the Nobel Prize in 1936.

[2] Del Rossi G., Malaquti A., and Del Rossi S. (2014). Practice effects associated with repeated assessment of a clinical test of reaction time. *Journal of Athletic Training* **49**(3), 356–359.

*Chi-Squared Data Analysis and Model Testing for Beginners.* Carey Witkov and Keith Zengel.
© Carey Witkov and Keith Zengel 2019. Published in 2019 by Oxford University Press.
DOI: 10.1093/oso/9780198847144.001.0001

**Table 1** *Falling meter stick data for the first five trials.*

| Displacement (m) |
| --- |
| 0.27 |
| 0.33 |
| 0.29 |
| 0.31 |
| 0.26 |

How would you use the displacement data to estimate your reaction time? One possibility is to average the displacements and substitute the mean displacement into Eq. 2. We could write this in equation form as

$$t^* = \sqrt{\frac{2\langle d \rangle}{g}},\tag{3}$$

where the angular brackets $\langle$ and $\rangle$ denote the mean value of the thing enclosed and an asterisk $t^*$ denotes the best estimate of $t$. Another possibility is to add a second column to Table 1 labeled Time (s), compute a reaction time for each displacement using Eq. 2, and average the reaction times. We could write this in equation form as

$$t^* = \left\langle \sqrt{\frac{2d}{g}} \right\rangle.\tag{4}$$

Would these two algorithms give the same results? If not, which one would give the correct result?

As both algorithms involve averaging it may be helpful to review why we average data in the first place. We average data because we believe that there is a "true" value and that the measurement uncertainty causes small, random deviations from this true value. These deviations are equally likely to be positive or negative, so averaging (which involves summing) will tend to cancel these positive and negative deviations from the true value.

So which algorithm, averaging displacements or averaging reaction times, is the "correct way" to estimate reaction time from displacement data? Well, averaging causes deviations from the true value to cancel when each deviation contributes equally and has an equal chance of being positive or negative. The equation for computing the reaction time (Eq. 2) applies a square root operation to the displacement measurements. The square root operation weights some displacements more than others, so the positive and negative deviations from the true value are weighted differently and can no longer be expected to cancel by simple averaging. Figure 3 shows the subtle shift caused in this case by this square root.

Therefore we should average the displacements in order to cancel these deviations. This is a rare case where the simpler of two algorithms, merely substituting once into the equation using mean displacement, is better! This can be stated in the form of a maxim: **Average the data, not the calculations!**

(a) Meter stick drop displacement data and mean displacement value.

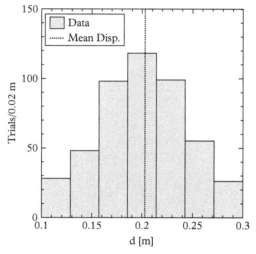

(b) Meter stick drop calculated time data, with mean values for the average-then-calculate and calculate-then-average algorithms shown.

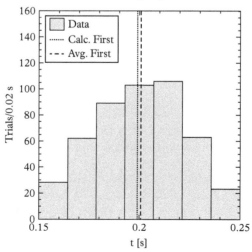

**Figure 3** *Histograms of the displacement (a) and time (b) for 500 simulated meter stick drops.*

## 2.2  Standard Deviation (RMSD)

Let's say you and your friend (or whoever) thought this calculated reaction time was an important scientific result. You write a paper on your experiment and results and send it to *The Journal of Kinesthetic Experiments (The JoKE)*, but it gets bounced back. You left something out. You forgot to include an estimate of uncertainty in your reaction time. The uncertainty in your reaction time depends on the spread (synonyms: dispersion, scatter, variation) in the data. How do you characterize the spread in data? Listing all of your data is usually difficult to read and impractical. Giving highs and lows is better than nothing (though not by much) as they disregard all of the other data and only characterize outliers. Taking the deviation (difference) of each value from the mean seems useful but runs into the problem that the mean deviation is zero.

To get around the problem of zero mean deviation, we have a few options. We could take the absolute value of the deviations before finding their mean. That would give a positive and non-zero value, but it's not fun working with absolute values in calculus. A second option is to square the deviations. This solves one problem but creates another. The mean squared deviation is positive but we now have square units (square meters in this case). To solve the square units problem we take the square root of the mean of the square deviations. So the algorithm is: calculate the deviations, square them, find their mean, and take the square root. If you read this backwards you have the root-mean-square deviation, or RMS deviation. In fact, you can re-create the algorithm for $N$ measurements of a variable $y$ by reading root-mean-square deviation word by word from right to left.

First, write the deviations: $y_i - \langle y \rangle$.
Second, square the deviations: $(y_i - \langle y \rangle)^2$.
Third, take the mean of the square of the deviations: $\frac{\sum (y_i - \langle y \rangle)^2}{N}$.
Fourth, take the square root: $\sqrt{\frac{\sum (y_i - \langle y \rangle)^2}{N}}$.

Of course, this result is better known as the standard deviation,[3] but it is really the root-mean-square deviation (RMSD),

$$\text{RMSD} = \sqrt{\frac{\sum_{i=1}^{N} (y_i - \langle y \rangle)^2}{N}}. \tag{5}$$

---

[3] You may find that in the standard deviation equation people sometimes use an $N$ in the denominator, while other people use an $N-1$. You may also find that people have long arguments about which is better or correct. We'll keep it simple and stick to the RMSD, which by definition has an $N$ in the denominator.

## 2.3  Standard Error

You resubmit your results to *The JoKE*, this time presenting your mean reaction displacement plus or minus its standard deviation, but your submission is bounced back again. The editor explains that the standard deviation does not measure uncertainty in your *mean* reaction displacement. Then what does standard deviation measure?

The standard deviation measures the spread of your entire set of measurements. In this experiment, the standard deviation of the displacements represents the uncertainty of each individual displacement measurement.

But the uncertainty of an individual measurement is not the uncertainty of the *mean* of the measurements. Consider the (mercifully) hypothetical case of taking a hundred measurements of your reaction time. If you repeated and replaced one of your measurements, you wouldn't be surprised if it was one standard deviation from the mean value of your original 100 measurements. But unless you were using a very long stick *and* something went terribly, medically wrong, you wouldn't expect the mean value of your new set of a hundred measurements to be very different from the mean value of your original hundred measurements. See figure 4 for an example of data from one hundred measurements, with and without the last trial repeated and replaced.

Now consider the hypothetical case of taking only four measurements, finding the mean and standard deviation, and then repeating and replacing one of those measurements. See figure 5 for an example of data from four measurements, with and without the last trial repeated and replaced. You would still find it perfectly natural if the new measurement were one standard deviation from the mean of your original four measurements, but you also wouldn't be so surprised or find much reason for concern for your health if your new mean value shifted by a considerable amount. In other words, the uncertainty of your next measurement (the standard deviation) won't change all that much as you take more data, but the uncertainty on the *mean* of your data should get smaller as you take more data. This is the same logic we used previously to explain why we averaged the displacement data: every measurement includes some fluctuation around the true value, which cancels out when we take the average. We don't expect these experimental uncertainties to be magically resolved before the next trial, but as we take more data, the mean value is less susceptible to being significantly shifted by the fluctuation in any one particular measurement.

The uncertainty of the mean (called the *standard error* or $\sigma$) should depend on the number of measurements (N), as well as on the spread (RMSD) of the data.

The equation for standard error is

$$\sigma = \frac{\text{RMSD}}{\sqrt{N}}. \tag{6}$$

For now we'll just state the result, but readers who are interested in further details should consult Problem 2.2.

(a) Meter stick drop displacement data and mean displacement value.

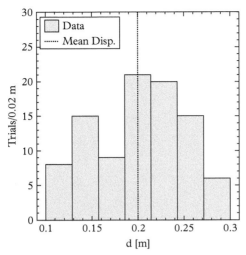

(b) Meter stick drop displacement data and mean displacement value for one hundred measurements, with a different measured value for the last trial.

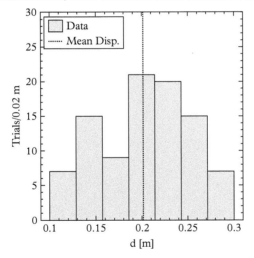

**Figure 4** *Histograms of the displacement for one hundred simulated meter stick drops (a) and for the same one hundred simulated meter stick drops with the last trial repeated and replaced (b). The shift in the mean value is barely noticeable, even though one measurement has moved from the far left bin to the far right bin in (b).*

(a) Meter stick drop displacement data and mean displacement value.

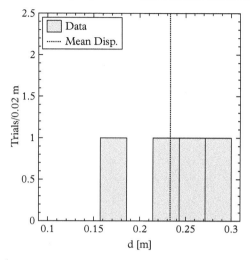

(b) Meter stick drop displacement data and mean displacement value, with a different measured value for the last trial.

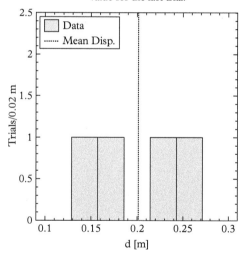

**Figure 5** *Meter stick drop displacement data and mean displacement value for four measurements, with different measured values for the last trial.*

## 2.4 Uncertainty Propagation

You resubmit your mean reaction displacement and its uncertainty, expressed as its standard error. Again, your manuscript bounces back!

You computed the standard deviation of the displacements using the RMSD algorithm. You divided this by the square root of the number of measurements to get the standard error of the mean of the displacements. The problem is that you still need to translate the spread in reaction displacements to a spread in reaction times. Equation 2 transformed the mean displacement to obtain the best estimate for a reaction time. Likewise, the displacement standard error must be transformed to obtain the best estimate for the uncertainty of the reaction time. The interesting thing about transformation of uncertainties in equations (called uncertainty propagation or error propagation, depending upon whether one is an optimist or a pessimist!) is that it isn't done simply by plugging the uncertainties into the equation.

The uncertainties we're considering are small differences, so we can obtain a formula for the propagation of the uncertainty in a variable $y$ through a function $f(y)$ using a little calculus:

$$\sigma_f = \left| \frac{\partial f}{\partial y} \right|_{\langle y \rangle} \sigma_y, \tag{7}$$

where $\langle y \rangle$ is the average value of your measurements of $y$ and the $\sigma$'s represent the uncertainty associated with each variable. The logic of this equation is that any curve over a small enough interval is roughly linear. Therefore if we want to scale a small spread in $y$ to a small spread in $f(y)$, as in figure 6, we should multiply by the slope at that point in the curve.

This equation allows us to use the uncertainty in a directly measured variable (the standard error of the displacement) to obtain the uncertainty of a calculated variable. When we apply Eq. 7 to the reaction time example we find

$$\sigma_t = \left| \sqrt{\frac{1}{2g\langle d \rangle}} \right| \sigma_d. \tag{8}$$

In this example, we only needed to propagate the uncertainty of one variable. For functions that depend on more than one variable ($f(x, y, z, ...)$), we can expand Eq. 7 to the multivariable case:

$$\sigma_f^2 = \left| \frac{\partial f}{\partial x} \right|^2 \sigma_x^2 + \left| \frac{\partial f}{\partial y} \right|^2 \sigma_y^2 + \left| \frac{\partial f}{\partial z} \right|^2 \sigma_z^2 + ..., \tag{9}$$

where the partial derivatives are evaluated at the mean values $\langle x \rangle$, $\langle y \rangle$, $\langle z \rangle$, etc. See Problem 2.3 if you're interested in learning how to derive this relationship.

You resubmit your manuscript one last time. This time you provide your best estimate reaction time and its uncertainty, correctly propagated from the standard

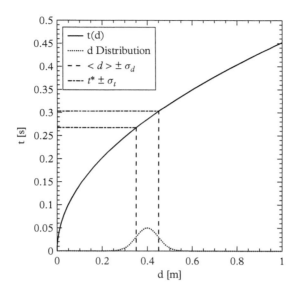

**Figure 6** *The transformation of the spread of measurements of displacement to a spread of times using the function* $t = \sqrt{2d/g}$. *Notice that the curve is roughly linear over this small region.*

error of the displacements. The editor finally agrees that your calculations are okay, but is concerned that your results will be of limited interest, even to the people who get *The JoKE*. Nevertheless, the editor publishes your results ... as a letter to the editor!

## 2.5   Random versus Systematic Uncertainties and Rounding Rules

In this book we will assume that all relevant uncertainties are random. As discussed earlier, averaging is a powerful technique because summing causes the small fluctuations around the true value to (mostly) cancel. However, this is only true if the fluctuations are actually random. It is possible that you might encounter systematic uncertainties in your experiment. Systematic uncertainties cause deviations from the true value that move in the same way or according to the same function in all your measurements. Common examples of systematic uncertainties are calibration shift, scale, and resolution uncertainties. A shift uncertainty occurs when the zero point of your measuring device is wrong. This could happen in a meter stick with a worn edge, such that it starts at the one millimeter mark, or with a scale that hasn't been zeroed and measures two grams with nothing on it. A scale uncertainty might occur if you have a budget meter stick with evenly spaced tick marks, even though the true length of the stick is 1.03 meters. A resolution uncertainty might occur if you are using a meter stick with marks only at every centimeter, or a voltmeter that has tick marks at every volt, when measuring something that varies at a smaller scale.

In general, systematic uncertainties are caused by issues related to your measuring devices. There are some advanced techniques for handling systematic uncertainties,[4] but we will not discuss them here. Instead, we will derive a methodology for experiments where random uncertainties dominate and are accurately described by the standard error on the mean, which can be calculated from repeated measurements.

Of course, to some extent, systematic uncertainties are present in any experiment. The total uncertainty is caused by the sum of the random and systematic uncertainties, which we know from propagation of uncertainty add in quadrature[5] (squared and summed) to create the total uncertainty

$$\sigma_{\text{total}}^2 = \sigma_{\text{random}}^2 + \sigma_{\text{systematic}}^2. \tag{10}$$

As you can see, the assumption of random uncertainty holds when $\sigma_{\text{random}}$ is larger than $\sigma_{\text{systematic}}$, and as you will see in Problem 2.6, it doesn't need to be all that much larger. On the other hand, this equation gives us a limit on the application of the standard error. Since the standard error decreases by a factor of $1/\sqrt{N}$, it is possible to take a large number of measurements and find a very small standard error. For example, some inexpensive force sensors are capable of taking over twenty samples per second. If you sampled with one of these sensors for a few minutes and then calculated the standard error, you could come up with an artificially low uncertainty estimate. How low is too low? Well, if you look closely at the actual values you are measuring, or if you plot a histogram of your measurements, you may find that the variation in each sample is really only the last digit of the force measurement bouncing around between two or three values (for example, if your measurements are all 5.12 or 5.13). That is because the variation in your measurement is very close to the resolution of your measuring device. In these cases, you should use the RMSD (standard deviation) for the uncertainty instead of the standard error, because the RMSD does not vary substantially with the number of measurements. If on the other hand your measuring device reports the exact same value every time, then your device is not precise enough for your measurements. The conservative way to handle such cases is to report the uncertainty as being a 5 in the place to the right after the last significant figure of your measured value. For example, 5.12 should have an uncertainty of $5.120 \pm 0.005$. The idea here is that your device is reporting the

---

[4] See, for example, Bechhoefer, J. (2000). Curve fits in the presence of random and systematic error, Am. J. Phys. **68**, 424.

[5] Sometimes you will see scientists report their random ("statistical") uncertainties and their systematic uncertainties separately. For example, you might see a reaction time of $(0.14 \pm 0.02 \text{ stat.} \pm 0.01 \text{ syst.})$ seconds. This is done so readers can determine the relative contributions to uncertainty, which helps in more sophisticated techniques of comparing results across experiments that have different systematic uncertainties.

most precise value it can, and that it would round up or down if the measurement were varying by more than this amount.

To summarize, the rules we propose for assessing uncertainties are:

1. Use the standard error by default.

2. If your measurement variations are on the order of the resolution of your device, use the RMSD.

3. If there is no variation in your measured value and no device with better precision is available, use a 5 in the place to the right of the last significant figure of your measurement.

Once you've established your uncertainty, you're ready to report your measurements. We recommend using the rounding rules given by the Particle Data Group,[6] who are responsible for maintaining a frequently updated list of nearly 40,000 experimental particle physics measurements. These experts use the "354 rule," which states that the nominal (mean) value should be rounded to the number of digits in the reported uncertainty, and the uncertainty follows these rounding rules:

1. If the three highest order digits in the uncertainty lie between 100 and 354, round to two significant digits.

2. If they lie between 355 and 949, round to one significant digit.

3. If they lie between 950 and 999, round up to 1000 and keep two significant digits.

For example, if your mean and standard error are $1.5947 \pm 0.0124$, you should report $1.595 \pm 0.012$. If your mean and standard error are $1.5947 \pm 0.0516$, you should report $1.59 \pm 0.05$. If your mean and standard error are $1.5947 \pm 0.0986$, you should report $1.59 \pm 0.10$.

## 2.6   Problems for Chapter 2

### 2.1   Average or calculate first?

If you have several measurements of a variable $y$, and you want to plug them into a function $f(y)$, then you should average first and then find $f(\langle y \rangle)$. How does $f(\langle y \rangle)$ differ from $\langle f(y) \rangle$? (Hint: Try Taylor expanding $f(y)$ around $\langle y \rangle$ to second order, then writing $\langle f(y) \rangle$ in terms of what you know about $y$.) How does this affect the result for the particular case of $t = \sqrt{2d/g}$?

---

[6] Patrignani, C. et al. (Particle Data Group) (2016). Chin. Phys. C. **40**.

## 2.2 Standard error derivation

One way to prove that the standard error on the mean is $\mathrm{RMSD}/\sqrt{N}$ is to use the propagation of uncertainty formula on the formula for the mean, $\langle y \rangle = \sum y_i/N$, where $\langle y \rangle$ is a function of every $y_i$, and treat the deviation from the mean of each value as an uncertainty: $\sigma_{y_i} = (y_i - \langle y \rangle)$. Use this logic to derive the formula for the standard error of the mean.

## 2.3 Correlated uncertainties

In deriving Eq. 9, we assumed that the deviations in the different variables were uncorrelated, that is, $\sigma_{xy} = \langle (x - \langle x \rangle)(y - \langle y \rangle) \rangle = 0$. Work out the case for a function of correlated variables, $f(x, y)$. (Hint: try Taylor expanding $f(x, y)$ to first order around $\langle x \rangle$ and $\langle y \rangle$, then calculating the RMSD of $f(x,y)$, $\langle [f(x,y) - f(\langle x \rangle, \langle y \rangle)]^2 \rangle$.)

## 2.4 Divide and conquer

Propagate the uncertainty for a ratio of two correlated variables, $f(x, y) = x/y$, where the corresponding uncertainties are $\sigma_x, \sigma_y$, and $\sigma_{xy}$ (defined in Problem 2.3). How does the uncertainty of the ratio compare to the individual uncertainties? Is this result useful for reporting experimental results?

## 2.5 Be fruitful and multiply

Propagate the uncertainty for a product of two correlated variables, $f(x, y) = xy$, where the corresponding uncertainties are $\sigma_x, \sigma_y$, and $\sigma_{xy}$ (defined in Problem 2.3). How does the uncertainty of the product compare to the individual uncertainties? Is this result useful for reporting experimental results?

## 2.6 Negligible uncertainties

Let's assume our measurement depends on the addition of two uncorrelated variables, $f(x, y) = x + y$, with corresponding uncertainties $\sigma_x$ and $\sigma_y$. Propagate the uncertainty and consider three cases: $\sigma_x = \sigma_y$, $\sigma_x = 5\sigma_y$, and $\sigma_x = 10\sigma_y$. At what point does the contribution to the total uncertainty from $\sigma_y$ become negligible?

# 3

# One Parameter Chi-squared Analysis

## 3.1 Modeling a System

To develop a model of a physical system, a physicist will work from a set of basic assumptions, like Newton's laws, to derive equations that explain the relationship between different measurable values. For example, a physicist could use Newton's second law to derive a model for the acceleration of a block down a frictionless incline,

$$a = g \sin \theta. \tag{11}$$

Any other physicist should be able to read this equation and comprehend that the acceleration of the sliding object depends on two things: the acceleration ($g$) due to gravity and the angle ($\theta$) of incline. The simplest experimental test of this model is to set up an incline, measure the sine of the angle, and measure the acceleration. The acceleration can be found by using the kinematic equation

$$x = \frac{1}{2}at^2 \tag{12}$$

and measured values of $x$ and $t$, which can be found with a motion sensor or a meter stick and stopwatch.

Suppose you did those measurements, and for a ramp of $\theta = 10°$, you found an acceleration of $a = 1.6 \pm 0.1$ m/s$^2$. You might be satisfied with that result. After all, we know that $g = 9.8$ m/s$^2$ and $g \sin 10° = 1.7$ m/s$^2$, which is perfectly consistent with your measurement. But is that enough to say that the $g \sin \theta$ model works? What if you showed this result to a skeptical friend who was convinced that this was just a coincidence? After all, you didn't try $20°$ or $50°$ or $87.8°$, did you? If the acceleration depends only on $g$ and the angle, shouldn't you have to show it for every angle to be truly convincing?[1] So you (begrudgingly) go back

---

[1] Let's assume your friend isn't so skeptical that they need you to do this experiment on different planets with different values of $g$!

*Chi-Squared Data Analysis and Model Testing for Beginners.* Carey Witkov and Keith Zengel.
© Carey Witkov and Keith Zengel 2019. Published in 2019 by Oxford University Press.
DOI: 10.1093/oso/9780198847144.001.0001

**Table 2** *Block on incline data.*

| Incline Angle | Acceleration [m/s$^2$] | Corresponding $g$ [m/s$^2$] |
|---|---|---|
| 10° | 1.6 ± 0.1 | 9.2 ± 0.6 |
| 20° | 3.3 ± 0.1 | 9.6 ± 0.3 |
| 30° | 4.9 ± 0.1 | 9.8 ± 1.0 |
| 40° | 6.4 ± 0.1 | 10.0 ± 0.2 |
| 50° | 7.3 ± 0.1 | 9.5 ± 0.1 |

and measure the acceleration for more angles, listed in Table 2. You share your results with the same friend, but now they're really confused. To them, it looks like you've measured several different values of $g$. They ask which value of $g$ they should believe, and how certain you are that your measurements are actually consistent. What can you tell them in reply?

Your friend is very insightful. What they are getting at is the idea that a model describes a relationship between different measurable physical quantities. There is a dependent variable ($a$ in this case) that will change if you change the independent variable ($\sin\theta$ in this case). Variables can take on different values (by definition), so in order to test a model it is important to test the relationship between the independent and dependent variables over a wide range of values. The fundamental question becomes, *what do your combined measurements tell you about your model?*

## 3.2 Uncertainties and the Central Limit Theorem

The first thing we need to understand and make peace with is that every measurement comes with an inherent uncertainty. Just as we saw in the reaction time experiment from Chapter 2, every time we repeat a measurement we get a different result. When we slide something down a low-friction incline, we still expect to get a different answer every time. Why? Well, for a lot of reasons. Literally. The table the incline is on is shaking just a little bit; there are air currents in the room that cause drag forces on the object; the ramp may rotate or tilt slightly between trials; the ramp is not perfectly frictionless and the coefficient of friction may not be uniform across the ramp; the stopwatch, meter stick, and motion sensors you used are not perfect, etc. All that, and we haven't even mentioned the possibility of gravitational waves or quantum effects! It is impossible to account for every little thing that goes into disrupting our measurements. From the vantage point of our experiment, the contributions from each of these sources are random. Randomness added to randomness added to randomness... and so on! This sounds like a dire situation. Physicists believe in *laws* of the universe, truths that are enforced and enacted everywhere and always. Yet they never measure the same thing twice. Never. The randomness seems insurmountable, incompatible with any argument for the laws that physicists describe.

Except it's not.

What saves physicists from this nihilism is the Central Limit Theorem. The Central Limit Theorem is one of the most perplexing and beautiful results in all of mathematics. It tells us that when random variables are added, randomness is not compounded, but rather reduced. An order emerges.

Formally, the Central Limit Theorem says that the distribution of a sum of random variables is well approximated by a Gaussian distribution (bell curve). The random variables being summed could have come from any distribution or set of distributions (uniform, exponential, triangular, etc.). The particular set of underlying distributions being summed doesn't matter because the Central Limit Theorem says you still get a Gaussian.[2]

One benefit of the Central Limit Theorem is that experimenters measuring a deterministic process that is influenced by many small random effects (such as vibrations, air currents, and your usual introductory physics experiment nuisances) need not be concerned with figuring out the distribution of each of these small random contributions. The net result of these random effects will be a Gaussian uncertainty.

A second benefit of the Central Limit Theorem is that the resulting distribution, a Gaussian, is simple to describe. The distribution of a sum of random variables will have a mean and a standard deviation as its dominant characteristics. The curve defined by these characteristics is the Gaussian,

$$G(x) = \frac{1}{\sqrt{2\pi}\sigma} e^{-(x-\mu)^2/2\sigma^2}. \tag{13}$$

The Gaussian is defined by two parameters: the mean ($\mu$, which shifts the curve as in figure 7a) and the standard deviation ($\sigma$, which makes the argument of the exponential dimensionless and defines the spread of the curve as in figure 7b). That's it. Sometimes people give other reasons why the Gaussian should be the curve that describes the uncertainties of our measurements (notably including Gauss himself). In our opinion, the most convincing reasoning is the Central Limit Theorem itself.

## 3.3   Using a Fit to Combine Measurements

The goal of a fit is to find the line that is closest to all of the points. In the case of a linear model with no intercept ($y(0) = 0$), the goal is to find the slope, $A$,

---

[2] If your jaw hasn't dropped yet, consider again how remarkable it is that, starting with random variables drawn from any distribution (even a flat uniform distribution), summing them, and plotting the distribution of the sums, you end up with a Gaussian!

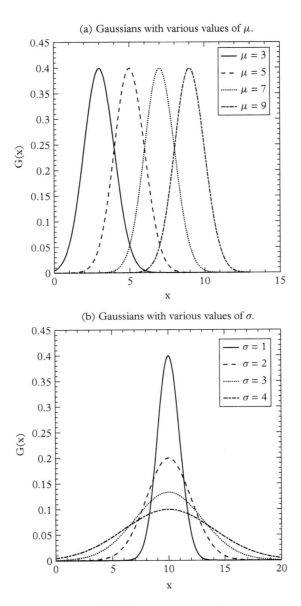

**Figure 7** *Examples of Gaussians with different parameter choices.*

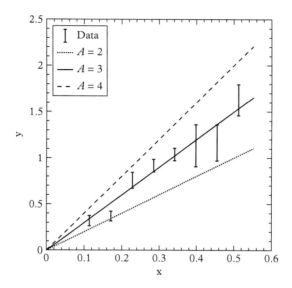

**Figure 8** *Three fit lines with different slopes for some arbitrary sample data.*

that accomplishes this.[3] See figure 8 for an example of some possible fit lines with different slopes for some sample data. To find the slope that gives the best fit we will need to define a cost function, a function that describes how close each point is to the fit line. Right now we are only interested in $y(x) = A x$ models where we can change $x$ and measure the resulting $y$.[4] We will assume that the uncertainties on $x$ are negligible compared to the uncertainties on $y$, so when we look for the distance of each point from the fit line, we will only consider the distance along the $y$-axis.

One approach would be to add up the differences between each measured point and each predicted point,

$$C(A) = \left| \sum_i (y_i - Ax_i) \right|. \tag{14}$$

A plot of these differences, also called residuals, is shown in figures 9a and 9b. The slope that minimizes this sum (has the lowest "cost") would then define the line of best fit. But there are a few problems with this method. Firstly, regardless of the data points, the minimum value of this sum will always be zero (see Problem 3.1). That means that all lines of best fit are judged as equally satisfactory

---

[3] We will discuss the case of models that have both a slope and an intercept in Chapter 4.

[4] If you're thinking that this type of model is overly simple and never comes up in real physics, see Problem 5.1 for examples of different physics formulas that can be re-cast in this simple linear form.

by this method. Secondly, half of the total residual differences of the points from the line of best fit will be positive and half negative. Because positive residuals cancel with negative residuals, the line of best fit could run right through the middle of all of the data points without being near any one of them! The worst case scenario associated with this would be a single outlier that draws the best fit line away from the rest of the data.

Clearly, a better approach is needed. One idea is that the problems with this method can be solved by defining a sum that has only positive contributions. One technique is to take the absolute values of each deviation,

$$C(A) = \sum_i |y_i - Ax_i|. \tag{15}$$

This solves one problem, but introduces another. Absolute values tend to be difficult to handle when dealing with calculus, which we would need in order to minimize this cost function. Another solution is to square each contribution,

$$C(A) = \sum_i (y_i - Ax_i)^2. \tag{16}$$

A plot of these differences is shown in figure 9c. This was Gauss' solution, which is now called least squares fitting (for obvious reasons). It is probably the most popular solution to the problem. After all, it takes care of the problem of opposite sign deviations canceling, it relies purely on a single measurement for each value of the independent variable, and the calculus needed to solve the minimization problem is straightforward. In fact, as a function of $A$, this cost function is a parabola, so you can find the minimum using only algebra! For almost 200 years, people were satisfied with this solution.

But we propose that it isn't good enough. We can do better.

While least squares is used everywhere, from scientific calculators to spreadsheet software, it still has some important drawbacks. First, it doesn't solve the outlier problem. A single data point that is not in line with the others can draw the line of best fit away from most of the data. Second, it is still somewhat arbitrary. Squares are easier than absolute values for the purposes of calculus, but squares and absolute values are not the only functions that always return positive values. If all we want are positive values, why don't we raise the deviations to the fourth power or the sixth or the $2n$-th? Why not use $\cos(y_i - Ax_i) + 2$? Third, least squares ignores crucial information about your measurements: the uncertainties. In least squares, every measured point is treated as having an exact location, which is of course a fantasy. Even worse, least squares weights each data point measurement exactly the same, even if you know that the uncertainty on some points is far larger than on other points.

One way to solve all of these problems is to use $\chi^2$ minimization.

(a) Theory curve and measured data, with residual differences shown.

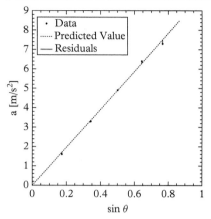

(b) Residual differences between theory and data.

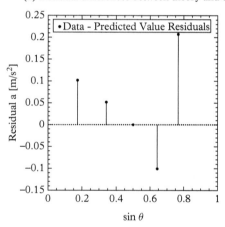

(c) Squared residual differences between theory and data.

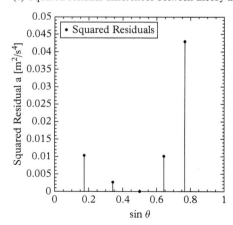

**Figure 9** *Plots showing the residual differences used in different cost functions for a g = 9.8 theory model line and the data shown in Table 2.*

## 3.4  Chi-squared Fitting

The $\chi^2$ (pronounced "kai squared") cost function is defined as

$$\chi^2 = \sum_i \frac{(y_{\text{measured } i} - y_{\text{model } i})^2}{\sigma_i^2}, \tag{17}$$

which in the case of a $y(x) = A\,x$ model is

$$\chi^2 = \sum_i \frac{(y_i - Ax_i)^2}{\sigma_i^2}, \tag{18}$$

where $\sigma_i$ is the uncertainty of the $i$th measurement. You might notice that this is a variation of least squares where the contribution from each measurement is weighted. Others have noticed this before, and this technique is typically called weighted least squares. But $\chi^2$ fitting is a particular version of weighted least squares where the weights are given by the uncertainties of the measurements. The cost function for weighted least squares has dimensions of $y^2$, whereas $\chi^2$ is dimensionless. So for the chi-squared version of weighted least squares, the line of best fit is the line that is the least number of sigmas away from the measurements. In other words, the deviations $(y_i - Ax_i)$ are put into units of uncertainty $(\sigma_i)$.

The problem of outliers is solved by $\chi^2$ fitting, since outliers can only substantially affect the line of best fit if they are particularly trustworthy, low uncertainty measurements. The definition of $\chi^2$ is also not at all arbitrary, though we'll defer discussion of the origins of $\chi^2$ until Section 3.5. The maths, though more complicated than regular least squares, is also still very manageable. Let's take a look.

The best fit value of the slope $A$ is the one that minimizes $\chi^2$. It is tempting to start taking derivatives to solve for the extrema, and second derivatives to examine whether the extrema are minima or maxima, but there is a better, more enlightening way. Let's start by expanding the square in the numerator of Eq. 18,

$$\chi^2 = \left[\sum_i \frac{y_i^2}{\sigma_i^2}\right] - 2A\left[\sum_i \frac{y_i x_i}{\sigma_i^2}\right] + A^2\left[\sum_i \frac{x_i^2}{\sigma_i^2}\right]. \tag{19}$$

The sums inside the square brackets are determined by the measurements, but they're still just numbers. For the sake of simplicity, we can give them different names,

$$\chi^2 = S_{yy} - 2AS_{xy} + A^2 S_{xx}, \tag{20}$$

where $S_{yy}$, $S_{xy}$, and $S_{xx}$ have replaced the sums. It should be obvious from Eq. 20 that $\chi^2$ is quadratic. That means it's a parabola! To solve for the minimum value of $\chi^2$ and the corresponding best fit $A$ value, we need only complete the square (see Problem 3.3) and rewrite $\chi^2$ as

$$\chi^2 = \frac{(A - A^*)^2}{\sigma_A^2} + \chi_{min}^2, \tag{21}$$

where $S_{yy}$, $S_{xy}$, and $S_{xx}$ have been exchanged for the three new parameters $A^*$, $\sigma_A$, and $\chi_{min}^2$. From algebra alone we can solve for the best fit value of $A$, $A^*$. The algebra will also tell us the minimum value of $\chi^2$, $\chi_{min}^2$, without resorting to any calculus. By this simple method we can find our best fit parameter $A^*$ and an estimate of how good a fit we have in $\chi_{min}^2$.

But how low should $\chi_{min}^2$ be? It's tempting to think zero is the optimal case, since after all we're looking for a minimum. But think of what it would mean to have a $\chi_{min}^2$ close to zero. One way would be to get extraordinarily lucky and have every point land exactly on the line of best fit. This is rare. In fact, you know it's rare because you know that there is uncertainty in every measurement. Every measurement landing exactly on the line of best fit is just as unlikely as measuring the exact same value several times in a row. Another way to have a very low $\chi_{min}^2$ is to have very large uncertainties, as the $\sigma$'s are in the denominator of $\chi^2$ in Eq. 18. If this happens, it might mean that you've overestimated your uncertainties. It might also mean that you have not varied your independent variable significantly between measurements (imagine measuring the acceleration of an object down an incline at $1.01°$, $1.02°$, $1.03°$ ...), or that your measuring device is simply not sensitive enough.

So you don't want $\chi_{min}^2$ to be close to zero. That doesn't mean you want it to be very large, either. You want it somewhere in the middle. A trustworthy value of $\chi_{min}^2$ is roughly the number of measurements you've taken. Loosely speaking, this would mean that every measurement is, on average, one $\sigma$ from the line of best fit, which is about as good as you can expect from an uncertain measurement. More rigorously, we can consider what the mean value of $\chi_{min}^2$ would be if we had the correct model and repeated the experiment (the entire set of $N$ measurements) several times,

$$\langle \chi^2 \rangle = \sum_{i=1}^{N} \frac{\langle (y_i - A_{\text{``true''}} x_i)^2 \rangle}{\sigma_i^2} = \sum_{i=1}^{N} \frac{\sigma_i^2}{\sigma_i^2} = \sum_{i=1}^{N} 1 = N, \tag{22}$$

where the angular brackets signify the mean value of everything enclosed. There are two tricks used in Eq. 22. First we have to assume that for the $i$th measurement, if $A_{\text{``true''}}$ is the correct model parameter value, then $\langle y_i \rangle = A_{\text{``true''}} x_i$. If that's true, then $\langle (y_i - A_{\text{``true''}} x_i)^2 \rangle$ is the mean squared deviation, which is the root mean square deviation, squared! In other words (or letters), $\langle (y_i - A_{\text{``true''}} x_i)^2 \rangle = \sigma_i^2$.

So far, we have seen that $\chi^2$ can be used to find the best fit, $A^*$. Further, we have a criterion for determining whether the best fit is a good fit: a good fit will have a similar value of $\chi^2_{min}$ as would the true model,

$$\chi^2_{min} \approx N \quad \text{for a good fit.} \tag{23}$$

Already, these chi-squared goodness of fit criteria are more informative than what is offered by least squares. Yet chi-squared analysis can tell us even more! To understand this extra information, we'll need to discuss a little bit of probability theory.

## 3.5 Maximum Likelihood Estimation

From the Central Limit Theorem we know the probability distribution function for most measurements is Gaussian. This means that the probability of measuring a value in a specific range is proportional to the magnitude of the Gaussian at that value ($p(y) \propto G(y)$). If we somehow had the correct model ($A_{\text{"true"}}x$) and the uncertainties, then we would know the probability of measuring any $y_i$ as long as we knew the corresponding $x_i$. Of course, we don't know the correct model—that's what we're trying to estimate. What we really want is to flip the question. We're not interested in how probable the data are given the model; we want to know how likely the model is given the data.[5] Notice the subtle switch in words in the previous sentence: data are "probable" while models are "likely". You may be tempted to ask, "what is the probability of this model?" But physicists tend to refrain from asking this question.[6] Physicists typically believe that there is a right answer, not a probability distribution of possible answers that sets real values in constant fluctuation (if the fundamental constants are actually changing over time, then energy is not conserved and we have some major problems!). So physicists don't ask which model is most probable. Instead they ask which model would make the data most probable, and they call that model the most likely model. Physicists are interested in the model of maximum likelihood.

To clarify the idea of probability versus likelihood, consider the example of dice rolling. What is the probability that you'll roll two ones in a row (snake eyes)? Well,

---

[5] Paraphrasing from JFK's inaugural speech, "Ask not, what data are consistent with my model; ask instead, what models are consistent with my data."

[6] A quick footnote for those familiar with probability theory: actually, physicists ask this question all the time. This is a Bayesian analysis, based on Bayes' theorem that $p(M|D)p(D) = p(D|M)p(M)$, where $p(D|M)$ means "the probability of the data given the model." We can eliminate $p(D)$ using the normalization condition (integrating both sides over $M$), but if we want to find $p(M|D)$ we'll still need to know $p(M)$, the absolute probability of a model. In other words, you can find the probability of the model given the data as long as you already know the probability of the model, which, as you can see, is sort of begging the question.

there's a 1/6 chance for the first roll and a 1/6 chance for the second roll. We're only interested in the second roll given that the first roll is a one, so to combine the probabilities we should multiply (note that the combined probability gets smaller with each extra specified outcome). That means there is a 1/36 chance of rolling snake eyes.

But we could ask a different question. What if we rolled the same die one hundred times and got a one thirty-eight times? You would be pretty sure that die was unfairly weighted. The likelihood of that die being fair (each of the six possible outcomes being equally likely) is low, because the probability that a fair die would produce that outcome is low.

Sometimes people find it helpful to distinguish probability and likelihood by thinking of probabilities as being characteristic of future outcomes, while likelihoods are characteristic of past outcomes. Typically, this is a useful distinction, but to avoid drawing unwanted criticism from the philosophically minded, we won't assert its absolute truth here.

In summary, we're interested in the most likely model, which is the model that makes the data most probable.

The first step is to write down the probability of our data given the unknown model ($y(x) = A\,x$). The likelihood is just that probability as a function of model parameter $A$. As we already know and as is shown in figure 10, the uncertainties of each point are Gaussian distributed.

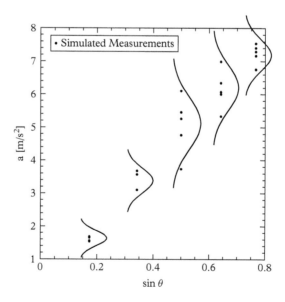

**Figure 10** *Simulated data for measurements of the acceleration of a block on a frictionless incline at different angles with the Gaussian uncertainty distributions overlaid.*

To find the likelihood, we'll multiply the probability of every measured value, which means we'll multiply on a Gaussian for every data point,

$$L(A) = p_{\text{total}} = p_1 \times p_2 \times p_3 \times \ldots \propto e^{\frac{-(y_1 - Ax_1)^2}{2\sigma_1^2}} \times e^{\frac{-(y_2 - Ax_2)^2}{2\sigma_2^2}} \times e^{\frac{-(y_3 - Ax_3)^2}{2\sigma_3^2}} \times \ldots \tag{24}$$

Note that the $\propto$ symbol means "is proportional to." Of course, when we multiply exponentials we add the arguments, which means

$$L(A) = p_{\text{total}} \propto e^{\frac{-1}{2} \sum_i \frac{(y_i - Ax_i)^2}{\sigma_i^2}} = e^{-\chi^2/2}. \tag{25}$$

This is exactly where chi-squared comes from: the combined Gaussian uncertainties of multiple measurements. Note also that the minimum value of $\chi^2$ corresponds to the maximum likelihood. That means that by solving for $\chi^2_{\text{min}}$ we've already solved the problem of maximum likelihood! We can plug Eq. 21 into Eq. 25 to find an even more illuminating result,

$$L(A) = e^{-\chi^2_{\text{min}}/2} e^{\frac{-(A - A^*)^2}{2\sigma_A^2}}. \tag{26}$$

The likelihood is (shockingly) a Gaussian as a function of $A$! Recall that we calculated $\sigma_A$ using algebra in Section 3.4. It should be obvious now why we chose to name this parameter as we did: $\sigma_A$ corresponds to the standard deviation of the Gaussian distribution of values of $A$.

Further, we know from Eq. 21 that $\chi^2_{\text{min}} + 1$ corresponds to an $A$ value of $A^* \pm \sigma_A$. This means that we can find the $A^* \pm \sigma_A$ range by inspecting the points where $\chi^2_{\text{min}} + 1$ intercepts the $\chi^2(A)$ parabola. An example of this is shown in figure 11. Note also that by the same logic, $\chi^2_{\text{min}} + 4$ corresponds to $A^* \pm 2\sigma_A$.

One important result for Gaussian distributed variables is that they fall within one sigma of the mean 68% of the time and within two sigmas of the mean 95% of the time. It is tempting to think that since your parameter $A$ is Gaussian distributed it must have a "68% probability" of falling within $A^* \pm \sigma_A$, but unfortunately this is not so. The real statement is about your measurements: if you repeated your entire experiment—all $N$ of your measurements—several times each and calculated a new $A^*$ each time, then 68% of those $A^*$s would fall within the original $A^* \pm \sigma_A$.

The true power of $\chi^2$ fitting (maximum likelihood estimation) is that it uses the Gaussian uncertainties of each of your measurements to find a Gaussian uncertainty on your best fit slope parameter. No other fitting algorithm does this. Least squares, for example, can give you a best fit, but $\chi^2$ fitting can give you a best fit $(A^*)$, a test for goodness of fit $(\chi^2_{\text{min}} \approx N)$, and an uncertainty on the best fit $(\sigma_A)$!

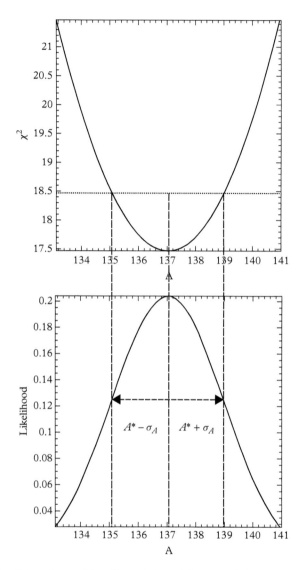

**Figure 11**  *The $A^* \pm \sigma_A$ band of the $\chi^2$ parabola and Gaussian likelihood.*

## 3.6   Example: Block on a Frictionless Incline

As an example, let's revisit the data in Table 2. We can replace $x_i$ and $y_i$ with $\sin \theta$ and $a$ in Eq. 19 to find

$$\chi^2 = \left[ \sum_i \frac{a_i^2}{\sigma_i^2} \right] - 2A \left[ \sum_i \frac{a_i \sin \theta_i}{\sigma_i^2} \right] + A^2 \left[ \sum_i \frac{\sin \theta_i^2}{\sigma_i^2} \right]. \tag{27}$$

Plugging in the values from Table 2, we find

$$\chi^2 = [139.7] - 2A[1356.2] + A^2[13171]. \tag{28}$$

You should confirm for yourself that $A^*$, $\sigma_A$, and $\chi^2_{\min}$ are 9.7, 0.8, and 5.4, respectively.

This is fake data, but the result is pretty good. The correct answer of $A^* = g = 9.81$ is just outside $A^* + \sigma_A$ and $\chi^2_{\min} \approx 5$, which means the fit is a good fit!

## 3.7  Problems for Chapter 3

### 3.1  Least absolute sum minimization

Prove that the minimum value (with respect to $A$) of $\left|\sum_i (y_i - Ax_i)\right|$ is zero.

### 3.2  Least squares minimization

Using calculus or algebra, solve for the value of $A$ that minimizes $\sum_i (y_i - Ax_i)^2$.

### 3.3  $\chi^2$ minimization

Solve for $A^*$, $\sigma_A$, and $\chi^2_{\min}$ in Eq. 21 in terms of $S_{yy}$, $S_{xy}$, and $S_{xx}$ in Eq. 20.

### 3.4  The RMSD of $\chi^2$

In this chapter we showed that $\chi^2_{\min} \approx N$ for a good fit, since $\langle \chi^2 \rangle = N$ the "true" model. What is the RMSD of $\chi^2$? (Hint: You will need to know that the mean quartic deviation, $\langle (x-\mu)^4 \rangle$, also known as the *kurtosis*, is $3\sigma^4$ for a Gaussian distributed random variable $x$. It might also be useful to know that the expected value of the product of two independent random variables is $\langle x \cdot y \rangle = \langle x \rangle \langle y \rangle$.)

Answer:

$$\mathrm{RMSD}(\chi^2) = \sqrt{2N}. \tag{29}$$

With this result we can better quantify goodness of fit, by examining whether $\chi^2_{\min}$ for a fit is within $N \pm \sqrt{2N}$.

### 3.5  More is better

Based on the result of the previous problem, what is the benefit of taking more measurements (data points)?

# 4

# Two–Parameter Chi-squared Analysis

In Chapter 3 we solved the problem of $\chi^2$ minimization for a one parameter (slope only) linear model. In this chapter, we will employ the same techniques to solve the problem of $\chi^2$ minimization for a two-parameter (slope and intercept) linear model. In general, $\chi^2$ minimization works for any model where $y$ is a function of $x$. It is simple enough to write down the equation for $\chi^2$:

$$\chi^2 = \sum_i \frac{(y_i - y(x_i))^2}{\sigma_i^2}. \tag{30}$$

A problem occurs when you try to estimate the parameters in $y(x)$ that minimize $\chi^2$. For the one-parameter linear model, $\chi^2$ is a parabola. For a $k$-parameter nonlinear model, $\chi^2$ can take on more complicated geometries, which makes it significantly more difficult to solve for the minimum. One solution to this problem is to transform your nonlinear model into a linear one.[1] We'll discuss how exactly this can be achieved in Chapter 6, but for now, let us assume that we can write any model that concerns us in terms of a two-parameter linear model ($y(x) = A x + B$).

Start by writing down $\chi^2$ for a two-parameter linear model:

$$\chi^2 = \sum_i \frac{(y_i - Ax_i - B)^2}{\sigma_i^2}. \tag{31}$$

We can expand the numerator to find

$$\chi^2 = \left[\sum_i \frac{y_i^2}{\sigma_i^2}\right] - 2A\left[\sum_i \frac{y_i x_i}{\sigma_i^2}\right] - 2B\left[\sum_i \frac{y_i}{\sigma_i^2}\right]$$
$$- 2AB\left[\sum_i \frac{x_i}{\sigma_i^2}\right] + A^2\left[\sum_i \frac{x_i^2}{\sigma_i^2}\right] + B^2\left[\sum_i \frac{1}{\sigma_i^2}\right]. \tag{32}$$

---

[1] See Problem 6.1 in Chapter 6 for examples.

*Chi-Squared Data Analysis and Model Testing for Beginners.* Carey Witkov and Keith Zengel.
© Carey Witkov and Keith Zengel 2019. Published in 2019 by Oxford University Press.
DOI: 10.1093/oso/9780198847144.001.0001

To save some writing, we can redefine those sums and rewrite this as

$$\chi^2 = [S_{yy}] - 2A[S_{xy}] - 2B[S_y] - 2AB[S_x] + A^2[S_{xx}] + B^2[S_0]. \qquad (33)$$

Just as in Chapter 3, we can complete the square (see Problems 3.3 and 4.1). This time, we'll be exchanging our six parameters (the $S$'s) for six new parameters, $A^*$, $\sigma_A$, $B^*$, $\sigma_B$, $\rho$, and $\chi^2_{min}$:

$$\chi^2 = \frac{(A-A^*)^2}{\sigma_A^2} + \frac{(B-B^*)^2}{\sigma_B^2} + 2\rho\frac{(A-A^*)(B-B^*)}{\sigma_A\sigma_B} + \chi^2_{min}. \qquad (34)$$

Of course, you recognize this as the equation of an elliptic paraboloid! In case you need a refresher, this means that in place of a parabola along the $A$ axis we now have a (bowl shaped) three-dimensional paraboloid hovering above the $A$–$B$ plane. The best fit parameters $A^*$ and $B^*$ are located at the minimum point of the paraboloid, where $\chi^2 = \chi^2_{min}$. We can also solve for the $\chi^2_{min} + 1$ and $\chi^2_{min} + 4$ (one and two $\sigma$) contours. In the one-dimensional case the $\chi^2_{min} + 1$ line intersects the $\chi^2$ parabola at exactly two points. In the two-dimensional case the $\chi^2_{min} + 1$ plane intersects the paraboloid at a set of points that fall on the elliptic cross-sections of the paraboloid (see figure 12).

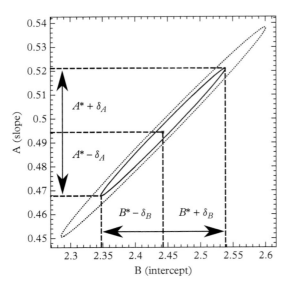

**Figure 12** *Contours where $\chi^2_{min} + 1$ and $\chi^2_{min} + 4$ interesect the hyperbolic paraboloid of $\chi^2(A, B)$ for arbitrary sample data.*

We are interested in the values of $A$ and $B$ that fall within the one-sigma contour. The easiest way to find each of these ranges is to project the edges of the ellipse onto the $A$ or $B$ axis. You might think that these correspond to $A^* \pm \sigma_A$ and $B^* \pm \sigma_B$, but the answer is slightly more complicated because of the third term in Eq. 34, which is sometimes called the correlation or covariance term. The constant $\rho$ determines the angle of the elliptic contours. This means that the choice of $A$ values that reside within the one-sigma contour is directly related to our choice of $B$. A little bit of thought and study of figure 13 should clarify why this is. Start with the line of best fit, $y = (A^*)x + B^*$. If you increase $B$, then you are raising the $y$-intercept of your line. To stay near your data, you need to decrease the slope of your line. Likewise, if you decrease $B$ you need to increase $A$ to keep your line near your data.

This is exactly why the range of $A$ and $B$ values that fall within the one-sigma contour is greater than $A^* \pm \sigma_A$ and $B^* \pm \sigma_B$: if we change both $A$ and $B$ at the same time then we find a larger range of each that fit within the one-sigma contour.

To solve for this range of $A$ values, we need to find the values of $A$ that correspond to the maximum and minimum edges of the ellipse along the $A$ axis. The equation for the ellipse is

$$\Delta\chi^2 = \chi^2 - \chi^2_{min}, \tag{35}$$

where $\chi^2$ is given by Eq. 34 and $\Delta\chi^2 = 1$ for the one-sigma contour. These points are defined as the points where the slope along the $B$ axis is zero.[2] Taking a partial derivative with respect to $B$ of Eq. 35 gives

$$\left| \frac{\partial(\Delta\chi^2)}{\partial B} \right|_{A_{\delta A}, B_{\delta A}} \quad 0 = \frac{2(B_{\delta A} - B^*)}{\sigma_B^2} + 2\rho \frac{(A_{\delta A} - A^*)}{\sigma_A \sigma_B}, \tag{36}$$

where $A_{\delta A}$ and $B_{\delta A}$ are the $A$ and $B$ values at the maximum and minimum of the ellipse along the $A$ axis. We can plug $A_{\delta A}$ and $B_{\delta A}$ into Eq. 35 and use Eq. 36 to find

$$(A_{\delta A} - A^*) = \pm\sigma_A \sqrt{\frac{\Delta\chi^2}{(1 - \rho^2)}}. \tag{37}$$

We can follow the same derivation for $B$ to find

$$(B_{\delta B} - B^*) = \pm\sigma_B \sqrt{\frac{\Delta\chi^2}{(1 - \rho^2)}}. \tag{38}$$

---

[2] A fancier way to say this is that the gradient of the paraboloid will only have an $\hat{A}$ component, or $\nabla(\chi^2) = \lambda\, \hat{A} + 0\, \hat{B}$, where $\lambda$ is some constant.

(a) The $\chi^2$ paraboloid with extreme values of $A$ and $B$ labeled.

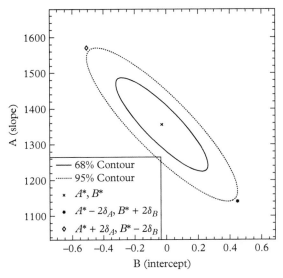

(b) Fit lines corresponding to the extreme values of $A$ and $B$ labeled in (a).

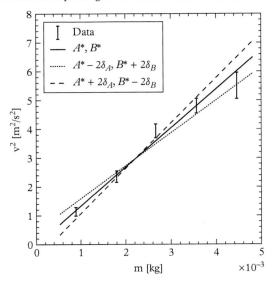

**Figure 13** *Effects of parameter variation on fit lines for arbitrary sample data.*

You may have noticed that in figure 13a the contours were labeled 68% and 95%, not $\chi^2_{min} + 1$ and $\chi^2_{min} + 4$. We choose to show the 68% and 95% contours to be consistent with the one-dimensional case, where $A^* \pm \sigma_A$ corresponds to the 68% interval and $A^* \pm 2\sigma_A$ to the 95% interval. The calculations are a bit more involved for the two-parameter model (see Section 7.2 in Chapter 7 if you're interested in the details), but for now let us simply assert that the relevant results are $\chi^2_{min} + 2.3$ and $\chi^2_{min} + 6.17$, which correspond to the 68% and 95% contours in the two-parameter case.

## 4.1   Problems for Chapter 4

### 4.1   Completing the square

Solve for $A^*$, $B^*$, $\sigma_A$, $\sigma_B$, $\rho$, and $\chi^2_{min}$ in Eq. 34 in terms of $S_{yy}$, $S_{xy}$, $S_y$, $S_x$, $S_{xx}$, and $S_0$ in Eq. 33. (Note: this problem will require plenty of algebra!)

### 4.2   Reference parameter values

Usually we don't have accurate estimates of A and B prior to an experiment because we lack accurate estimates of the quantities needed to calculate parameters A and B. Occasionally, however, we have the luxury of knowing accurate reference values for parameters A and B (imagine an experiment where parameters A and B are determined by combinations of well-known values such as $\pi$ or $g$). Explain how plotting known reference values for parameters A and B onto the contour plot generated by a two-parameter linear model chi-squared analysis can be used to test the model.

# 5

# Case Study 1: Falling Chains

## 5.1 Falling Chain Model

Let's use chi-squared analysis to test a model and obtain a best fit estimate of a model parameter. If a chain is dropped onto a scale, the scale reading will increase over time and approach a maximum value, then drop to the total weight of the chain. What is the ratio of the maximum force to the weight of the chain?

First, let's define some constants and variables: $y =$ the distance between the top link of the chain and the scale; $L =$ total length of chain; and $\lambda = \frac{dm}{dy} = \frac{M}{L} =$ density of chain. Figure 14 shows a diagram of the setup.

The force read by the scale is the result of two forces: the weight of the chain resting on the scale and the force of the chain colliding with the scale (impulse/time),

$$F_{scale} = F_{weight} + F_{impact}. \tag{39}$$

We can write $F_{weight}$ as

$$F_{weight} = mg = \lambda(L - y)g. \tag{40}$$

To find $F_{impact}$, consider the impulse the scale imparts on each link,

$$F_{s \to l}\, dt = dm\, \Delta v = dm\,(0 - v) = -v\, dm, \tag{41}$$

where $F_{s \to l}$ is the force of the scale on a single link and $dm$ is the mass of a link. Rearranging and using Newton's third law,

$$F_{l \to s} = v\frac{dm}{dt}. \tag{42}$$

But we know that

$$\frac{dm}{dt} = \frac{d(\lambda y)}{dt} = \lambda v. \tag{43}$$

*Chi-Squared Data Analysis and Model Testing for Beginners.* Carey Witkov and Keith Zengel.
© Carey Witkov and Keith Zengel 2019. Published in 2019 by Oxford University Press.
DOI: 10.1093/oso/9780198847144.001.0001

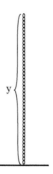

**Figure 14** *Sketch of a chain with mass M, total length L, in its starting position of y = L.*

With this we see that

$$F_{\text{scale}} = F_{\text{weight}} + F_{\text{impact}} = \lambda(L - y)g + \lambda v^2. \tag{44}$$

From one-dimensional kinematics (or alternatively the conservation of energy for each link) we know

$$v^2 = 2a(L - y), \tag{45}$$

which we can substitute into Eq. 44 to obtain

$$F_{\text{scale}} = 3\lambda(L - y)g. \tag{46}$$

Just as the last link of the chain reaches the scale, $y = 0$, so

$$F_{\text{max}} = 3\lambda Lg = 3Mg. \tag{47}$$

The maximum force of the chain on the scale can be written in the form of a one-parameter linear model $y = Ax$, where

$$y = F_{max}, \tag{48}$$

$$x = Mg, \tag{49}$$

and

$$A = 3. \tag{50}$$

Therefore the prediction is, when dropping a chain on a scale, the maximum force will be three times the weight of the whole chain.

## 5.2 Data Collection

To perform a chi-squared analysis, we need to vary something (the *x*-variable) and something else has to change in response (the *y*-variable). In this example we need to vary the mass of the chain. Instead of using multiple chains, each with a different mass, it's possible to use one chain, "varying" its mass by zeroing the scale with different amounts of chain resting on it before dropping the chain.

Recall[1] from Section 3.5 that chi-squared analysis assumes a Gaussian distribution of uncertainties around each data point. The central limit theorem ensures this only if data points are sums of random variables, e.g. mean values. Therefore, the "data points" (the points that are plotted) in a chi-squared analysis are not individual measurements but mean values. Recall also from Section 3.4 that chi-squared analysis requires uncertainties, which for mean values are described by their standard errors. The conclusion to draw from these two recollections about the need for mean values and standard errors in chi-squared analysis is that the chain must be dropped several times at one mass just to obtain one mean value and one standard error (just one data point and its uncertainty!) in the chi-squared analysis. To balance between an optimal result and a realistic time commitment, we recommend five trials (chain drops) for each measurement (chain mass).

Data recorded by a force plate (or a force sensor connected to a bucket) is often very noisy and the maximum force may appear as just one sampled value that varies a lot with repeated trials. One way to improve both the accuracy (true value) and precision (spread of results) is to recognize that the shape of the force curve versus time is quadratic (i.e. a parabola). We can prove this as follows.

The net force of the chain on the scale at any time is

$$F_{net} = 3\lambda(L - y)g. \tag{51}$$

Substituting the one-dimensional constant acceleration kinematic equation, $y = L - \frac{1}{2}gt^2$, gives

$$F_{net} = \frac{3}{2}\lambda g^2 t^2. \tag{52}$$

Therefore, the force of the chain on the scale increases quadratically with time until it reaches a maximum. By fitting the force curve[2] to a quadratic we can take the intersection of the quadratic curve with the time that the maximum force occurs.

A sample data set of maximum force measurements for different masses (of the same chain) is shown in Table 3.

---

[1] Often in this chapter you will be asked to "recall" something. If you don't recall the fact being referenced, please refer back to previous chapters to review.

[2] Here we can use a least squares fitting algorithm, as with a typical sensor we will have tens of measurements every second, each of which we want to contribute equally to the fit.

**Table 3** *Falling chain data.*

|  | Trial 1 | Trial 2 | Trial 3 | Trial 4 | Trial 5 |
|---|---|---|---|---|---|
| Mass [g] | $F_{max}$ [N] | $F_{max}$ [N] | $F_{max}$ [N] | $F_{max}$ [N] | $F_{max}$ [N] |
| 0.114 | 0.187 | 0.233 | 0.510 | 0.314 | 0.357 |
| 0.171 | 0.310 | 0.312 | 0.237 | 0.498 | 0.489 |
| 0.228 | 0.552 | 1.060 | 0.731 | 0.804 | 0.640 |
| 0.285 | 0.888 | 1.009 | 0.921 | 1.087 | 0.680 |
| 0.341 | 1.044 | 0.995 | 1.117 | 1.219 | 0.818 |
| 0.398 | 1.249 | 1.855 | 0.694 | 0.604 | 1.266 |
| 0.455 | 0.760 | 1.792 | 0.745 | 1.182 | 1.342 |
| 0.512 | 1.781 | 1.812 | 0.962 | 1.870 | 1.710 |

## 5.3 Chi-squared Analysis

As our model is a one-parameter linear model, we will use the one-parameter (1d) chi-squared analysis script. To run the script, we will need to populate the $x$-variable arrays with the chain weights and the $y$-variable arrays with the five $F_{max}$ values that correspond to each chain weight.

After filling in the arrays and running the one-parameter chi-squared script, the data plot shown in figure 15 and the chi-squared plot shown in figure 16 are obtained. Recall from Section 3.4 that the best fit is a good fit if $\chi^2_{min}$ is of the order of N, the number of data points. Since the chi-squared distribution has a mean of N and an RMSD of $\sqrt{2N}$, any value within $N \pm \sqrt{2N}$ is a good value of $\chi^2_{min}$. For eight data points, the $\chi^2_{min}$ value of around 10.5 is evidence of a good fit. A test of the model is the first result from chi-squared analysis as it wouldn't make sense to study the best fit parameters if the best fit wasn't a good fit.

Recall from Section 3.5 that Gaussian variables fall within one sigma of the mean 68% of the time. Thus if the best fit line is the correct model about which each point is randomly distributed, then we expect 68% or roughly 2/3 of the points to be within one sigma of the best fit line. We see that this is the case in figure 15, where six of the eight points are within one sigma of the best fit line.

Since the fit in this example appears to be good, we turn next to the parameter estimate $A_{best}$. Here we find that the best fit parameter $A$ is 2.96. That is, we find that the highest likelihood of obtaining the data occurred with the model $y = 2.96x$, compared to any other value of $A$ in the one-parameter linear model family of $Y = Ax$. Parameter estimation ($A^* = 2.96$) is the second result from our chi-squared analysis. The 2.96 result seems close to the predicted value of 3 from our theoretical model.

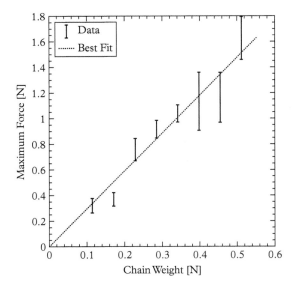

**Figure 15** *Falling chain data plot.*

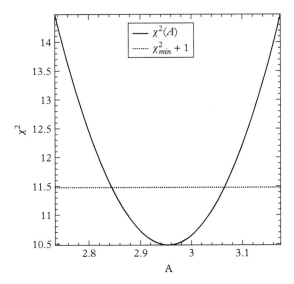

**Figure 16** *Falling chain $\chi^2$ plot.*

The next step is to obtain an uncertainty estimate for the best fit parameter. Recall from Section 3.4 that the $\pm$ one sigma lower and upper bounds on the best fit parameter correspond to $\chi^2_{min} \pm 1$. From the script results and from inspection of figure 16 we see that $\sigma_A = 0.11$. The best fit parameter uncertainty is the third result we have obtained from our chi-squared analysis. As we have the benefit of a model-predicted parameter value ($A = 3$) we can check whether it falls within one or two standard errors of the best fit parameter estimate ($A^*$ in the book and *Abest* in the scripts found in the Appendices). In this case the model-predicted parameter value of 3 falls within the uncertainty in $A^*$ so the model survives the test. The model would have been rejected if the model-predicted parameter value was more than two standard deviations[3] from the best fit parameter value. Model testing by parameter value is the fourth result obtained by our chi-squared analysis.

   This first case study involved a simple application of chi-squared analysis to perform model testing and parameter estimation along with parameter uncertainty. In addition to employing the chi-squared goodness of fit criterion, ($\chi^2_{min}$) should be of the order of $N$, the number of data points, this case study demonstrated use of chi-squared analysis' second model testing method that checks if reference parameter values (when available) are within two standard errors (two sigma) of best fit parameter values. However, there are still more features of chi-squared analysis not brought out in this case study. We didn't have the opportunity to revise our model and check if the revised $\chi^2_{min}$ value is closer to $N$, the number of data points, since the first model was a good fit. Also, this case study involved only a one parameter (slope only) linear model, not a full two parameter (slope and $y$-intercept) linear model with $\chi^2$ contour plots. Our second case study will provide an opportunity for model testing and parameter estimation with uncertainties in a two-parameter linear model and will demonstrate model revision and retesting, i.e. a complete chi-squared analysis!

## 5.4   Problems for Chapter 5

### 5.1   Nonlinear one parameter models

Write each of the following models in a linear form ($y = Ax$) such that the parameter is part of the slope, $A$: $n(t) = e^{-t/\tau}$, $h(r) = \ln(r/r_0)$, $v(d) = d/\sqrt{z}$, and $\omega(\Omega) = \beta\Omega/(1 + \beta)$.

---

[3] There are different rules in different disciplines for just how many sigmas warrant model rejection, but for now we'll use a two sigma rule.

# 6

# Case Study 2: Modeling Air Resistance on Falling Coffee Filters

Our first case study applied chi-squared model testing and parameter estimation to a system using a one-parameter (1d) model ($y = Ax$). As the model fit was good in the first case study no attempt was made to revise the model.

Our second case study is a more complete chi-squared analysis that includes one-parameter models, two-parameter models, and model revision. The second case study also demonstrates a new feature of chi-squared analysis: the ability to test a whole family of models.

In this case study we will model the air drag (a force) that acts on falling coffee filters. If you drop a basket-type paper coffee filter, it will accelerate for a little while, then essentially stop accelerating as it approaches its constant maximum speed, known as the terminal velocity. From Newton's laws we know that at terminal velocity there must be some upward force acting on the filter that cancels with the force of gravity. The upward force on the coffee filter is the force of air resistance, or air drag.[1] If you think about it, the drag force must depend on velocity. After all, the filter starts with zero velocity when you drop it, then speeds up for a while, then the drag force cancels with the weight once the filter reaches a certain velocity. The only way this can happen is if the drag force depends on velocity. But does the drag force depend linearly on $v$, or is the relationship quadratic or logarithmic or something even more exotic?

Let's develop a physical model for the drag force on the coffee filter. The air could be sliding along the sides of the coffee filter or pushing against the bottom of the coffee filter or both. Let's consider the sliding case first, collect some data, and test the model.

---

[1] We should all be thankful that air drag exists—it's the reason rain drops don't kill!

*Chi-Squared Data Analysis and Model Testing for Beginners.* Carey Witkov and Keith Zengel.
© Carey Witkov and Keith Zengel 2019. Published in 2019 by Oxford University Press.
DOI: 10.1093/oso/9780198847144.001.0001

## 6.1  Viscous Drag Model

A reasonable guess for the factors that determine the force of air sliding past the coffee filter are the area of the walls of the coffee filter (more area results in more sliding), the viscosity (the characteristic impulse per area from sliding) of the air, and the velocity gradient of the air around the filter (which characterizes how fast layers of air are moving relative to each other and how far apart they are from each other).

Combining these effects, we find

$$F = A\mu \frac{\Delta v}{\Delta r} = (2\pi rh)(\mu)\left(\frac{v}{\Delta r}\right), \tag{53}$$

where $r$ is the radius of the circular bottom of the filter, $h$ is the height of the walls of the filter, $\Delta r$ is the thickness of the layer of air that is moving with the filter due to friction, and viscosity is denoted by $\mu$, with units of Pascal-seconds. Figure 17 shows a sketch of this model of a falling filter and the surrounding air.

At terminal velocity the upward viscous drag force equals the coffee filter's weight,

$$mg = (2\pi rh\mu)v_T/\Delta r. \tag{54}$$

We can rewrite this equation into a one-parameter linear model ($y = Ax$) suitable for chi-squared analysis, where

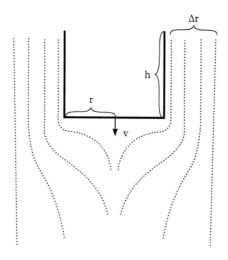

**Figure 17** *Sketch of a filter (solid) falling through layers of air (dotted). Note that the filter has been modeled as a cylinder for simplicity. Layers close to the filter are dragged along with it, while layers at $\Delta r$ and beyond are unaffected.*

$$v_T = \left( \frac{g \Delta r}{2\pi \, rh\mu} \right) m \tag{55}$$

becomes

$$y = Ax. \tag{56}$$

The mass of the coffee filters can be varied by stacking filters. Chi-squared analysis requires uncertainties, so we need many terminal velocity measurements at each mass. We also need enough different masses to sensibly test the model.[2] A reasonable trade-off between necessity and patience is five terminal velocity measurements at each of five different masses.

## 6.2 Data Collection

In our introductory physics lab course at Harvard, students use ultrasonic motion sensors pointing downward from a height of a little over two meters to collect displacement versus time data for falling coffee filters. To obtain terminal velocity data our students perform linear curve fits[3] to the displacement data in the region of terminal velocity. The best fit slope is the best estimate for the terminal velocity.[4]

Table 4 provides sample data obtained in a falling coffee filter experiment performed by a group of students in one of our labs.

Table 4 *Falling coffee filter data.*

|          | Trial 1      | Trial 2      | Trial 3      | Trial 4      | Trial 5      |
|----------|--------------|--------------|--------------|--------------|--------------|
| Mass [g] | $v_T$ [m/s]  | $v_T$ [m/s]  | $v_T$ [m/s]  | $v_T$ [m/s]  | $v_T$ [m/s]  |
| 0.89     | 1.150        | 1.026        | 1.014        | 1.134        | 1.010        |
| 1.80     | 1.443        | 1.492        | 1.583        | 1.563        | 1.589        |
| 2.67     | 2.031        | 1.939        | 1.941        | 2.067        | 1.941        |
| 3.57     | 2.181        | 2.202        | 2.199        | 2.109        | 2.269        |
| 4.46     | 2.507        | 2.292        | 2.303        | 2.364        | 2.267        |

---

[2] Paraphrasing Abraham Maslow, "if all you have are two points, the whole world looks like a line."
[3] Again, least squares is fine here, where we will have tens of measurements every second.
[4] Fitting is preferable to numerical differentiation because fitting averages by nature and numerical differentiation (calculating $\Delta y / \Delta x$) amplifies noise.

## 6.3   Chi-squared Analysis

To run the one-parameter chi-squared script, we need to populate the *x*-variable arrays with the filter masses and the *y*-variable arrays with the five terminal velocity measurements that correspond to each filter mass.

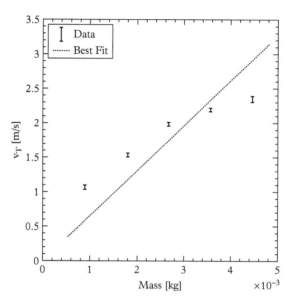

**Figure 18**  *Viscous drag model data and line of best fit.*

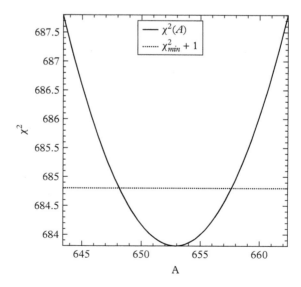

**Figure 19**  *Viscous drag model $\chi^2$ plot.*

Figure 18 shows the coffee filter data with error bars and the best fit line for the viscous drag model. We found in Chapter 1 that you can't rely on looking at a plot without zooming in with errors bars to see if a fit is good, but you can sometimes look at a plot and immediately tell if a fit is bad ... and this fit is bad!

Figure 19 shows the one-parameter chi-squared plot.

Recall from Section 3.4 that the best fit is a good fit if $\chi^2_{min} \approx N$. Since the chi-squared distribution has a mean of $N$ and an RMSD of $\sqrt{2N}$, any value within $N \pm \sqrt{2N}$ is a good value of $\chi^2_{min}$. Here, $\chi^2_{min}$ is 683.8 for five data points—ouch! We can safely reject the viscous drag model of falling coffee filters.

## 6.4 Revising the Model

With the rejection of the viscous drag (sliding friction) model it's time to take up the inertial drag (pushing friction) model.

In keeping with our goal of introducing chi-squared analysis to as wide a group of readers as possible, we'll develop an inertial drag model with and without calculus.

Here we will assume that the filter is pushing a thin region of air just beneath it. Figure 20 shows a sketch of this model.

Let's start the calculus-based model derivation by defining some variables:

$v$ = speed of the coffee filter,

$A$ = area of coffee filter,

$R$ = force of air resistance (drag) on coffee filter,

$\rho$ = mass density of air,

$dh$ = the height of a thin region of air that is in contact with the coffee filter in each $dt$,

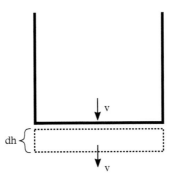

**Figure 20** *Sketch of a filter (solid) pushing a thin layer of air (dotted) of height dh.*

$dm$ = the mass of the thin region of air beneath the coffee filter, and

$dV$ = the volume of the thin region of air beneath the coffee filter.     (57)

Then,

$$R = \frac{dp}{dt} = m\frac{dv}{dt} + v\frac{dm}{dt}. \tag{58}$$

At terminal velocity $dv/dt = 0$, so

$$R = v\frac{dm}{dt}. \tag{59}$$

But

$$dm = \rho\, dV = \rho A\, dh, \tag{60}$$

so

$$R = v\rho A\frac{dh}{dt} = \rho A v^2. \tag{61}$$

The inertial drag model predicts that the force depends upon velocity squared, a quadratic velocity dependence.

A similar conclusion can be reached algebraically using dimensional analysis. The first step is to guess the variables that the force depends on, and raise them to unknown exponents and multiply them,

$$R = \rho^a A^b v^c. \tag{62}$$

The second step is to reduce each variable to a combination of the fundamental dimensions in mechanics: T (time), L (length), and M (mass),

$$M\frac{L}{T^2} = \left(\frac{M}{L^3}\right)^a \left(L^2\right)^b \left(\frac{L}{T}\right)^c. \tag{63}$$

The third step is to write exponent equations for each of the fundamental dimensions and solve for the unknown exponents,

$$M : 1 = a.$$

$$T : -2 = -c, \text{ so } c = 2. \tag{64}$$

$$L : 1 = -3a + 2b + c, \text{ so } 2 = 2b, \text{ or } b = 1.$$

The fourth and final step is to substitute the now known exponents back into the original equation to arrive at the dimensionally derived equation,

$$R = \rho A v^2. \tag{65}$$

Whichever method you used to arrive at this result, at terminal velocity, $R = mg$, so we have

$$mg = \rho \pi r^2 v_T^2. \tag{66}$$

Since we are only considering uncertainties in the $y$-variable, we should rewrite this as

$$v_T^2 = \frac{1}{\rho \pi r^2} mg. \tag{67}$$

We can view the preceding equation as a one-parameter linear model, $y = Ax$,

$$y = v_T^2, \tag{68}$$

$$A = \frac{1}{\rho \pi r^2}, \tag{69}$$

and

$$x = m. \tag{70}$$

In revising the model from viscous drag to inertial drag the following changes are needed in the one-parameter chi-squared script in the Appendix:

$$y = v^2 \tag{71}$$

and

$$y\_err = 2v\sigma_v. \tag{72}$$

The new $y\_err$ was obtained using the one-variable uncertainty propagation rule.

## 6.5   Testing the Revised Model

The resulting fit and $\chi^2$ parabola are shown in figure 21 and figure 22. After making these changes, we note that $\chi^2_{min}$ drops from its previous value of 683.8

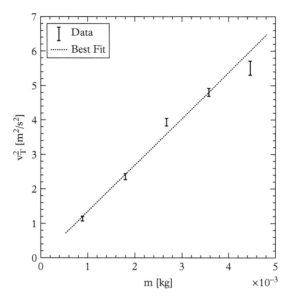

**Figure 21** *Inertial drag model data and line of best fit.*

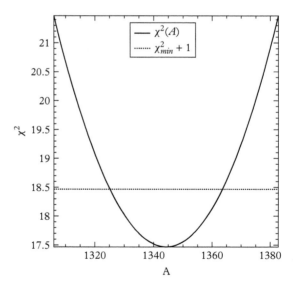

**Figure 22** *Inertial drag model $\chi^2$ plot.*

down to 17.5, suggesting that the revised model is a big improvement over the original model. Recall from Problem 3.4 that a good fit, where data points are about sigma from the model, results in $\chi^2_{min}$ falling within $N \pm \sqrt{2N}$, or in this case, within the range $5 \pm 3$. The $\chi^2_{min}$ value of 17.5 falls outside this range so the fit cannot be considered to be good despite being a vast improvement over the original model.

## 6.6   Finding the Optimal Model for a Dataset

So far we've seen that the inertial air drag model $(v^2)$ does a better job predicting the data than the viscous air drag model $(v^1)$. This may be because the true model has a $v^2$ dependence. Of course, all we've seen so far is that it is closer to the data. For all we know, the fit was better because the true model is *really* $v^3$ or $v^4$, and a $v^2$ model is closer to those than is a $v^1$ model. Now, we could test each one of these models by changing our $y$ and $y\_err$ values and finding a line of best fit for each of these models, but there's a better way: we can use chi-squared analysis to test a whole family of $v^n$ models!

First, we need to rewrite the inertial drag model in a form where the velocity exponent $n$ appears as one of the two parameters in a linear model $y = Ax + b$. This can be done as follows.

At terminal velocity,

$$R = mg. \tag{73}$$

We can assume that air drag is proportional to velocity to some power $n$, giving

$$kv^n = mg, \tag{74}$$

where $k$ is an unknown coefficient. We can take the natural logarithm of both sides to linearize this equation into a two-parameter linear model by writing

$$\ln(kv^n) = \ln(mg), \tag{75}$$

which reduces to

$$n\ln(v) + \ln(k) = \ln(mg). \tag{76}$$

This can be rearranged to find

$$\ln(v) = \frac{1}{n}\ln(mg) - \frac{1}{n}\ln(k). \tag{77}$$

We can write this model in the form

$$y = Ax + b, \tag{78}$$

where

$$y = \ln(v), \tag{79}$$

$$x = \ln(mg), \tag{80}$$

$$A = \frac{1}{n}, \tag{81}$$

and

$$B = -\frac{1}{n}\ln(k). \tag{82}$$

The best fit parameter of $A$ tells us the best velocity exponent for predicting our dataset, while the best fit parameter of $B$ includes many contributions and is more difficult to directly interpret. Again, a few modifications are needed in the two-parameter $\chi^2$ script to accommodate the new model:

$$y = \ln(v), \tag{83}$$

$$x = \ln(mg), \tag{84}$$

and

$$y\_err = \sigma_v/v. \tag{85}$$

The new $y\_err$ equation was again obtained using the one-variable uncertainty propagation rule. Figure 23 shows the data and line of best fit, while figure 24 shows the 68% and 95% contour lines centered on the best fit parameter values for $A$ and $B$. The $\chi^2$ script gives an $A^* \pm \delta_A$ value of 0.494 ± 0.027, which is consistent with the predicted value of $1/n = 0.5$. Here $\chi^2_{min} = 18$, even though there were only five data points. While the fit isn't within the good fit criterion, we will argue in Section 6.7 that outright rejection is not warranted.[5]

If we had the luxury of a reference value for parameter $B$ we could use the second model testing method that chi-squared provides, that is to check if the model-predicted parameters fall within the 95% contour line. Unfortunately, too many uncertain quantities compose $B$ (e.g. the drag coefficient based on the shape

---

[5] If you are thinking that a more complicated model is warranted, then you may want to skip ahead to Problem 6.2 for further discussion.

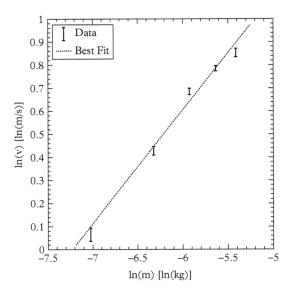

**Figure 23** *Data and best fit plot for $v^n$ model.*

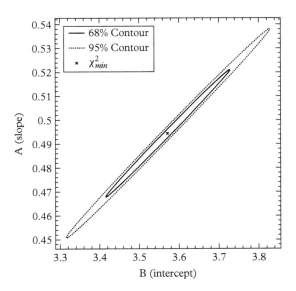

**Figure 24** *Contour plot for $v^n$ model.*

of the coffee filter), so while we have a model-predicted value for parameter $A$ ($A_{theory} = 0.5$, corresponding to $n = 2$), we have no model-predicted value for parameter $B$ and must rely solely on the relationship between $\chi^2_{min}$ and $N$ for model testing.

## 6.7   Using Chi-squared to Debug Experimental Issues

Of course, we still have the problem of a large $\chi^2_{min}$, which seems to suggest that we have the wrong model, even though we tested a whole family of velocity dependent models. Something suspicious is happening here, since the $v^n$ test told us that $n = 2$ is the best fit, and we've already seen what appears to be a reasonable fit in figures 21 and 22.

Let's take a close look at the plots showing the data and lines of best fit (Figures 21 and 23). In both cases we notice that there seems to be one point that is particularly far from the line of best fit. In fact, we can calculate the independent $\chi^2$ of each of these points, which can be seen in Table 5. The culprit seems to be the third measurement. In fact, we can see from figure 25, where this measurement has been excluded, that the fit appears to be unaffected. We can confirm this by inspecting figures 22 and 25, where we find that the two slopes are within two sigma of one another. Further, the $\chi^2_{min}$ is now 4.2, which is very much within the $N \pm \sqrt{2N} = 4.0 \pm 2.8$ range predicted.

The point here is not that you should start excluding points to produce a better fit: the point is that you can use $\chi^2$ analysis to discover *experimental* errors. The data presented here is real classroom data, obtained by students, and subject to experimental errors. We're talking now about errors (not just uncertainties), where something is procedurally done *wrong*. All but one of the data points are on a line and the best fit value ($n \approx 2$) matches with our expectation that we'll have an integer value exponent in our velocity dependence (not to mention the fact that we can look up the velocity dependence of the drag force on a coffee filter in certain textbooks!). This is a case where the multiple features of $\chi^2$ model testing can help you figure out what went wrong. Here it seems that something experimental was done differently with the third measurement from the other four. Perhaps the filters used in the third measurement were a little worn and therefore had a different shape. Maybe a door was opened, or people were walking by just then,

**Table 5** *Falling coffee filter data.*

|  | Measurement 1 $\chi^2$ | Measurement 2 $\chi^2$ | Measurement 3 $\chi^2$ | Measurement 4 $\chi^2$ | Measurement 5 $\chi^2$ |
|---|---|---|---|---|---|
| $v^2$ Model | 0.7841 | 0.5791 | 10.2653 | 0.0024 | 5.8355 |
| $\ln(v)$ Model | 1.3910 | 1.0626 | 9.8202 | 0.0049 | 5.3943 |

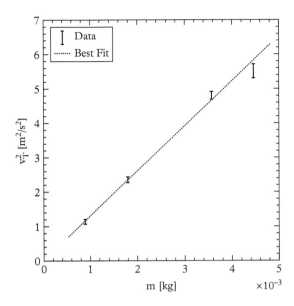

**Figure 25** *Inertial drag model data and line of best fit, with the third measurement excluded.*

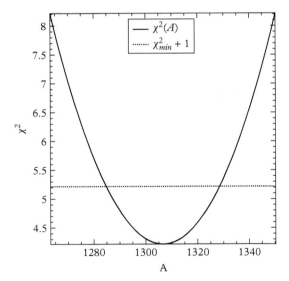

**Figure 26** *Inertial drag model $\chi^2$ with the third measurement excluded.*

or the students changed up their data collection routine somehow. Whatever the cause of this discrepancy, $\chi^2$ can help us figure out where things went wrong and what we can do next. This is the mark of a good model testing method: it not only reveals discrepancies between model and data, but also helps the experimenter decide whether the model or the data is at fault. Usually, and in this case, the thing to do is to revisit your experimental procedure, control for more confounding variables, and take more data!

## 6.8   Problems for Chapter 6

### 6.1   Nonlinear two-parameter models

Write each of the following models in a linear form ($y = Ax + B$) such that the parameters are part of the slope ($A$) or intercept ($B$): $E(t) = \sin(wt + \delta)$, $\phi(\theta) = 1/\sqrt{\alpha + \beta \cos \theta}$, and $R(v) = Av^2 + Bv$.

### 6.2   Calculating the Reynolds number

If you want to check whether an additional model revision can bring $\chi^2_{min}$ closer to $N$ for the falling coffee filter data, try combining the viscous and inertia drag models ($R = Av^2 + Bv$) and test. Using your results, estimate the ratio of inertial to viscous drag forces. This ratio is called the Reynolds number, and is used in fluid mechanics to classify different regimes of drag force behavior.

# 7

# Advanced Topics

In this chapter we present topics that are relevant to but not necessary for understanding chi-squared analysis. Each section here is independent. You can skip this chapter altogether if you like, or you can skip ahead to the sections that interest you.

## 7.1 Probability Density Functions

Probability density functions (pdfs) describe the probability that the outcome of some random process will have a value in some specific interval. To understand pdfs, let's start with a simple example. Say a teacher gives their class a test that is graded on a scale from one to ten. The teacher is particularly mean and unforgiving, so there is no partial credit. The grades in the class, in no particular order, are $9, 8, 9, 5, 7, 0, 9, 6, 7, 10$, and $8$. One question we can ask is, what is the average grade? Of course, this is a simple problem. You just add up all the scores and divide by the total number of students,

$$\text{Average} = \frac{(9 + 8 + 9 + 5 + 7 + 0 + 9 + 6 + 7 + 10 + 8)}{11} = 7.1. \tag{86}$$

We could generalize this formula to the case where we have $N$ students, each receiving a grade of $g_i$:

$$\langle g \rangle = \frac{1}{N} \sum_{i=1}^{N} g_i, \tag{87}$$

where $\langle g \rangle$ is the average or expected value of $g$. We can simplify this even further by using a histogram. A histogram is a plot that shows the number of events that correspond to each outcome. Here, the "events" are the students (taking the test) and the outcomes are their grades. A histogram of the grades for this particular example is shown in figure 27.

*Chi-Squared Data Analysis and Model Testing for Beginners*. Carey Witkov and Keith Zengel.
© Carey Witkov and Keith Zengel 2019. Published in 2019 by Oxford University Press.
DOI: 10.1093/oso/9780198847144.001.0001

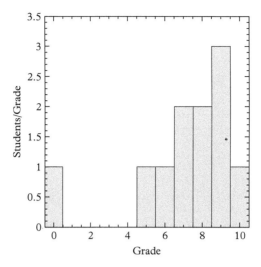

**Figure 27** *Histogram for eleven student grades.*

We can think of this figure as telling us that there are $n_i$ students who received each grade, $g_i$. In this case, we can write our average as

$$\langle g \rangle = \frac{1}{N} \sum_{i=1}^{N_{\text{bins}}} n_i g_i, \tag{88}$$

where

$$N = \sum_{i=1}^{N_{\text{bins}}} n_i. \tag{89}$$

If you like, you can think of this as a "weighted average," where grades are weighted by their frequency and then summed. For eleven students, this procedure doesn't save a whole lot of time. But in a class of 100 or 1000 students, this technique could save a lot of time! figure 28 shows histograms for 100 and 1000 students.

Next, let's imagine that the teacher's heart is not so hardened and they allow for half credit on each problem. How would we need to change our histograms? Well, we would need twice as many bins since there are twice as many possible grades (See figure 29). Following this trend, we could imagine a teacher who allows for partial credit down to the nearest 0.1 point on each problem. As you can see in figure 30, we'd need even more bins!

You may have noticed that we could have very smooth continuity between bins if we had two things: lots of bins and lots of students. As figure 30 shows, we could draw a curve that roughly matches a distribution of scores for 10,000 students

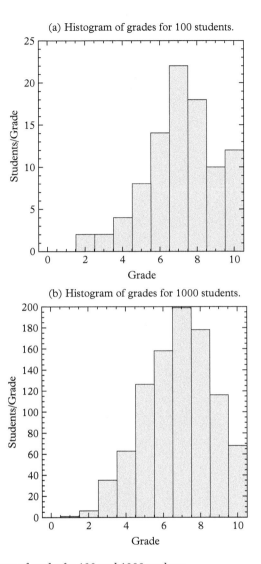

**Figure 28** *Histograms of grades for 100 and 1000 students.*

across 100 possible grades. The curve that matches the data in this plot is the pdf, the distribution of the probability density of different grades.

With a little bit of careful thought, we can write down the continuous analog of each element of Eqs. 88 and 89:

1. $g_i$ is the location on the $x$-axis.
2. $n_i$ is the location on the $y$-axis that corresponds to $g_i$.

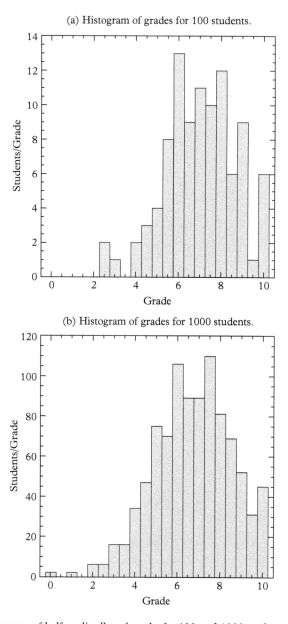

**Figure 29** *Histograms of half-credit allowed grades for 100 and 1000 students.*

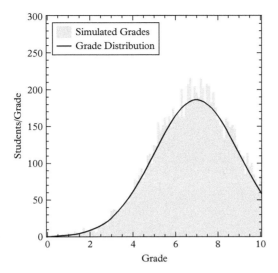

**Figure 30** *Histogram of 10,000 student grades over bins of 100 possible grades.*

3. The discrete sum, $\sum_{i=1}^{N_{bins}}$, becomes the continuous sum,[1] $\int_{g_{min}}^{g_{max}} dg$.
4. The bins become "bins" of $[g, g + dg]$.

Putting all of this together gives

$$\langle g \rangle = \frac{1}{N} \int_{g_{min}}^{g_{max}} n(g)g \, dg, \tag{90}$$

where

$$N = \int_{g_{min}}^{g_{max}} n(g) \, dg. \tag{91}$$

The associated jargon is $n(g)$ the probability density function (pdf), $g$ is the independent variable, and $1/N$ is the normalization constant.[2]

Usually, you'll see $p(x)$ and $x$, not $n(g)$ and $g$, and people integrate over all possible values of $x$:

$$\langle x \rangle = \frac{1}{N} \int_{-\infty}^{\infty} p(x)x \, dx. \tag{92}$$

---

[1] Yes, the $\int$ in integrals is a big S as in "$\int$um."
[2] The normalization constant is often included in the definition of the pdf, but we're keeping it separate here for emphasis.

We can also find the average value of any function of $x$, $A(x)$, with the following formula:

$$\langle A(x) \rangle = \frac{1}{N} \int_{-\infty}^{\infty} A(x)p(x)\,dx, \tag{93}$$

where $\langle A(x) \rangle$ is often referred to as the "expected value" of $A$, sometimes written as $E[A]$.

One famous and important expected value is the variance, which is the mean squared deviation of the distribution,

$$\mathrm{Var}[x] = \langle (x - \langle x \rangle)^2 \rangle = \frac{1}{N} \int_{-\infty}^{\infty} (x - \langle x \rangle)^2 p(x)\,dx. \tag{94}$$

With a little bit of algebra (see Problem 7.2), one can show that

$$\mathrm{Var}[x] = \langle (x - \langle x \rangle)^2 \rangle = \langle x^2 \rangle - \langle x \rangle^2. \tag{95}$$

Sometimes people experience a strong temptation to refer to $p(x)$ as "the probability of $x$," but it just isn't so; $p(x)$ is a probability density, which has units of probability per unit-$x$. Although you can't state the probability of some exact value of $x$ (think of how unlikely it is to measure an *exact* value with perfect precision down to any number of decimals), you can find the probability that $x$ falls in some range:

$$p(x_2 > x > x_1) = \frac{1}{N} \int_{x_1}^{x_2} p(x)\,dx. \tag{96}$$

A popular example of this is a $p$-value, which is the probability of finding a specified value of $x$ or greater:

$$p_{x_1} = p(x > x_1) = \frac{1}{N} \int_{x_1}^{\infty} p(x)\,dx. \tag{97}$$

## 7.2 The Chi-squared Probability Density Function

In Section 3.4 we showed that the mean value of a $\chi^2$ is (for a correct model) $N$, the number of data points. In Problem 3.5 you showed that the variance of $\chi^2$ is $2N$. For most introductory applications the mean and variance are sufficient for assessing the goodness of fit. For a more precise statement, it is sometimes useful to give the $p$-value of your minimum $\chi^2$, or in other words the probability that you would have obtained a value equal to or greater than your minimum $\chi^2$, assuming you had the correct model. To give a $p$-value, you would first need to know the probability density function of $\chi^2$.

The simplest way to solve for the $\chi^2$ pdf is to use a change of variables. Let's start with a single measurement, $y$. We know from the central limit theorem that $y$ is Gaussian distributed:

$$\int p(y)\, dy = \int G(y)\, dy = 1,\tag{98}$$

where $G(y)$ is the Guassian pdf. We want to go from $p(y)\, dy$ to $p(\chi^2)\, d(\chi^2)$:

$$\int_{-\infty}^{\infty} G(y)\, dy = 2\int_{0}^{\infty} G(y)\, dy = \int_{0}^{\infty} p(\chi^2)\, d(\chi^2) = 1.\tag{99}$$

The first equality in this equation holds because $G(y)$ is even, so $\int_{0}^{\infty} G(y)\, dy = \int_{-\infty}^{0} G(y)\, dy$. We want to rewrite our integral solely in terms of positive values of $y$ because we want the limits of the integral to match those of the $p(\chi^2)$ integral, which must run from 0 to $\infty$ because $\chi^2$ is always positive. The second equality is our definition of the thing we're looking for, $p(\chi^2)$. We can write $y(\chi^2)$ by rearranging $\chi^2$ for a single measurement,

$$\chi^2 = \frac{(y-\mu)^2}{\sigma^2},\tag{100}$$

to find

$$y = \sigma\sqrt{\chi^2} + \mu\tag{101}$$

and

$$\frac{dy}{d(\chi^2)} = \frac{\sigma}{2\sqrt{\chi^2}}.\tag{102}$$

Using Eq. 102, we can rewrite Eq. 99,

$$\int_{0}^{\infty} p(\chi^2)\, d(\chi^2) = 2\int_{0}^{\infty} G(\chi^2)\frac{dy}{d(\chi^2)}\, d(\chi^2) = \int_{0}^{\infty} \frac{2}{\sqrt{2\pi}\sigma}e^{-\chi^2/2}\frac{\sigma}{2\sqrt{\chi^2}}\, d(\chi^2),\tag{103}$$

from which we see

$$p(\chi^2) = \frac{1}{\sqrt{2\pi}}\frac{1}{\sqrt{\chi^2}}e^{-\chi^2/2}.\tag{104}$$

Things are a little different for two measurements, $x$ and $y$. The combined pdf for these two measurements is

$$G(x, y) = G(x)G(y) = \frac{1}{2\pi\sigma_x\sigma_y} e^{-\frac{(x-\mu_x)^2}{2\sigma_x^2} - \frac{(y-\mu_y)^2}{2\sigma_y^2}}. \tag{105}$$

We're interested in contours of equal probability in $G(x, y)$, which are given by

$$\frac{(x-\mu_x)^2}{\sigma_x^2} + \frac{(y-\mu_y)^2}{\sigma_y^2} = \chi_x^2 + \chi_y^2 = \text{constant}. \tag{106}$$

This means that we are interested in how the probability changes when we move to different contours of equal probability. The trick is to write our probability in terms of

$$\chi = \sqrt{\chi_x^2 + \chi_y^2} \tag{107}$$

using polar coordinates, which means that we must first change variables from $x$ and $y$ to $\chi_x$ and $\chi_y$:

$$\int\int G(x)G(y)\,dx\,dy = \int\int G(x)G(y)\frac{dx}{d\chi_x}\,d\chi_x\frac{dy}{d\chi_y}\,d\chi_y$$

$$= \int\int \frac{1}{2\pi} e^{-\chi_x^2/2 + \chi_y^2/2}\,d\chi_x\,d\chi_y. \tag{108}$$

We can introduce an angular coordinate, $\phi$, and rewrite this integral as

$$\int\int \frac{1}{2\pi} e^{-\chi^2/2}\chi\,d(\chi)\,d\phi = \int e^{-\chi^2/2}\chi\,d(\chi). \tag{109}$$

If this looks too fancy for you, consider that we're really just using calculus to convert from Cartesian to polar coordinates. Finally, we can change variables from $\chi$ to $\chi^2$ using $d\chi = d\chi^2/(2\chi)$, to find

$$p(\chi^2) = \frac{1}{2}e^{-\chi^2/2}. \tag{110}$$

We can find the pdf of $\chi^2$ for $N$ measurements by adapting the same technique. The combined pdf is given by multiplying all the Gaussians together,

$$\int G(y_1, y_2, y_3, ...)\,dy_1\,dy_2\,dy_3... = \int \prod_{i=1}^{N} G(y_i)\,dy_i = \int \prod_{i=1}^{N} \frac{1}{2\pi\sigma_i}e^{-(y_i-\mu_i)/2\sigma_i}\,dy_i. \tag{111}$$

We can rewrite this in terms of $\chi_i$ to find

$$\int \prod_{i=1}^{N} G(y_i)\, dy_i = \int \prod_{i=1}^{N} G(\chi_i) \frac{dy_i}{d\chi_i}\, d\chi_i = \int \prod_{i=1}^{N} \frac{1}{\sqrt{2\pi}} e^{-\chi_i^2/2}\, d\chi_i. \qquad (112)$$

The contours of constant probability are given by

$$\chi = \sqrt{\sum_i \chi_i^2} = \text{constant}, \qquad (113)$$

where $\chi$ is the radius of an $n$-sphere. The differential volume element can be written as

$$\prod_{i=1}^{N} dy_i = \chi^{N-1}\, d\chi\, dS_N, \qquad (114)$$

where $dS_N$ contains all the angular information about the $n$-sphere. We can rewrite Eq. 112 as

$$\int \prod_{i=1}^{N} \frac{1}{\sqrt{2\pi}} e^{-\chi_i^2/2}\, d\chi_i = \frac{1}{(2\pi)^{N/2}} \int dS_N \int e^{-\chi^2/2} \chi^{N-1}\, d\chi. \qquad (115)$$

Using

$$\int dS_N = \frac{2\pi^{N/2}}{(\frac{N}{2}-1)!} \qquad (116)$$

and $d\chi = d\chi^2/(2\chi)$, we find

$$p(\chi^2) = \frac{1}{2^{N/2}(\frac{N}{2}-1)!} e^{-\chi^2/2} (\chi^2)^{(N-2)/2}. \qquad (117)$$

For odd values of $N$, you will need to find the factorial of a non-integer number. This can be achieved with the gamma function,

$$\Gamma(N) = (N-1)!. \qquad (118)$$

The $\chi^2$ pdfs for various values of $N$ are shown in figure 31.

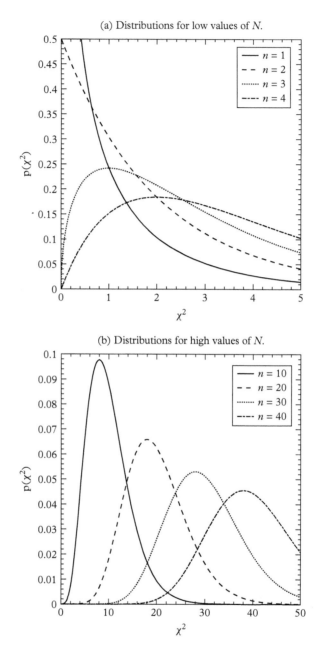

**Figure 31** *Distributions for various values of N.*

## 7.3 The Reduced Chi-squared and Degrees of Freedom

Sometimes you will see people refer to the "reduced chi-squared" values instead of the chi-squared values. The reduced chi-squared is just the regular chi-squared divided by the degrees of freedom, $\nu$:

$$\chi^2_{red} = \frac{\chi^2}{\nu}. \tag{119}$$

The number of degrees of freedom is defined as the number of measurements (data points on your plot) minus the number of parameters in your model. In our case, we have only considered one-parameter (slope only) and two-parameter (slope and intercept) models, so we have $\nu = N - 1$ or $\nu = N - 2$.

The reason people subtract out the number of model parameters is that they are interested in $\chi^2_{min}$, not $\langle \chi^2 \rangle$. We expect $\langle \chi^2 \rangle = N$ for $N$ measurements, as we showed in Eq. 22. But that is what we expect if we know the *true* model. The *best fit* model, on the other hand, is the one that minimizes $\chi^2$, and the minimization criteria are given by $A = A^*$ and $B = B^*$, both of which also depend on the data. Therefore the total degrees of freedom in $\chi^2_{min}$ is the number of measurements minus the number of constraints those measurements place on the best fit parameters.

One intuitive way to understand this concept is to imagine fitting a one-parameter model to a single measurement. What value of $\chi^2_{min}$ do you expect? Well, for a single point you can always find a slope that will send a line straight through the mean value of your measurement, so you should expect $\chi^2_{min} = 0$. For two measurements, you can again be sure that the best fit could match one point exactly, so you're really only testing the deviation of one of those measurements. For $N$ measurements, you're always getting one of them for free, because a line can match one of them exactly.

The same logic goes for two-parameter fits, where $\chi^2_{min} = 0$ for any experiment with only two measurements, since the two mean values are the two points you need to define a line.

When reporting results, you should make sure to include the number of measurements and the number of model parameters along with your minimum chi-squared or reduced chi-squared value. This way a reader can easily interpret your result without having to count data points or scour the text for your model definition!

## 7.4 The 68% and 95% Contours for Two Parameters

At the end of Chapter 4, we mentioned that for a two-parameter linear model, the $\chi^2_{min} + 1$ and $\chi^2_{min} + 4$ contours do not reflect one- and two-sigma deviations of the parameters. Instead, we chose to show the 68% and 95% contours in order to show something comparable to the one- and two-sigma deviations of the slope parameter (which is Gaussian) in the one-parameter model. We know that for two parameters we can write $\chi^2$ in the form of Eq. 34, and that the likelihood will be $e^{-\chi^2/2}$:

$$L(A, B) = e^{-\frac{(A-A^*)^2}{2\sigma_A^2} - \frac{(B-B^*)^2}{2\sigma_B^2} - \rho \frac{(A-A^*)(B-B^*)}{\sigma_A \sigma_B} - \frac{\chi^2_{\min}}{2}}. \tag{120}$$

This equation is not a Gaussian like the one-dimensional case, but it is another type of function that you are now familiar with, if you read Section 7.3. To make this clear, let's rewrite it as

$$L(C, D) = e^{-\frac{(C-C^*)^2}{2\sigma_C^2} - \frac{(D-D^*)^2}{2\sigma_D^2} - \frac{\chi^2_{\min}}{2}}, \tag{121}$$

where

$$C = \frac{A}{\sigma_A} + \frac{B}{\sigma_B},$$

$$C^* = \frac{A^*}{\sigma_A} + \frac{B^*}{\sigma_B},$$

$$\sigma_C = \sqrt{\frac{2}{1+\rho}},$$

$$D = \frac{A}{\sigma_A} - \frac{B}{\sigma_B}, \tag{122}$$

$$D^* = \frac{A^*}{\sigma_A} - \frac{B^*}{\sigma_B}, \text{ and}$$

$$\sigma_D = \sqrt{\frac{2}{1-\rho}}.$$

Here we've exchanged our *A*s and *B*s for *C*s and *D*s. This choice of variables may seem obscure, but really all we've done is chosen a new set of coordinates, *C* and *D*, that are oriented along the paraboloid, as shown in figure 32.

The important thing is that we have eliminated the pesky cross-term ($\rho(A - A^*)(B - B^*)/\sigma_A \sigma_B$ in Eq. 120. We went through all this trouble because the likelihood as written in Eq. 121 is just the $\chi^2$ pdf (see Eq. 110) for $N = 2$ measurements. This may seem a little weird, since the likelihood of the two parameters follows a $\chi^2$ pdf for two "measurements," no matter how many measurements you actually took. But it's not weird. In the likelihood, your measured values and uncertainties are all just constants. Only the parameters can vary, and there are only two parameters. All we've shown in this section so far is

(a) The 68% contour of the likelihood in the *A*–*B* coordinate system.

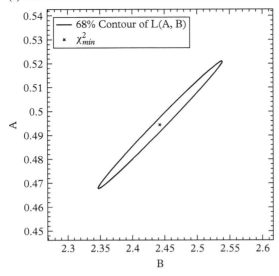

(b) The 68% contour of the likelihood in the *C*–*D* coordinate system.

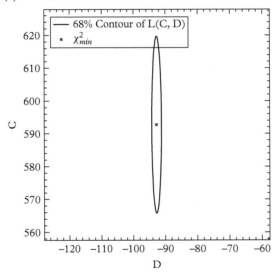

**Figure 32** *The 68% contours of the likelihood in different coordinate systems.*

that those two parameters are themselves $\chi^2$ distributed. Anyway, once we know how the parameters are distributed we can figure out where the 68% and 95% contours are. All we have to do is find $\chi^2_{68}$, the value of $\chi^2$ that encloses 68% of the total probability, which we can write as

$$0.68 = \int_0^{\chi_{68}^2} p(\chi^2)\, d(\chi^2) = \int_0^{\chi_{68}^2} \frac{1}{2} e^{-\chi^2/2}\, d(\chi^2). \tag{123}$$

We can follow the same procedure for the 95% contour.

In statistics jargon, the integral of a pdf, $p(x)$, up to a value of $x$ is called the cumulative distribution function (cdf), or $c(x)$. What we've just done is solve for the point where $c(\chi^2) = 0.68$.

In our case, the answers are $\chi_{68}^2 = 2.3$ and $\chi_{95}^2 = 6.2$, so we're interested in $\chi_{min}^2 + 2.3$ and $\chi_{min}^2 + 6.2$.

## 7.5  Poisson Distributed Uncertainties

Although we know from the central limit theorem that any repeated measurement which depends on the sum of several random sources of uncertainty will be Gaussian distributed, it is still true that there are special cases where this is not the case and therefore Gaussian uncertainties do not apply. One important class of exceptions is systems with Poisson distributed uncertainties.[3] Poisson uncertainties apply for measurements that take on only integer values. For example, the number of hurricanes that will land in the USA in any given year is an integer value (for better or worse there are no partial hurricanes) and Poisson distributed. Another famous example is the number of expected nuclear decays of a radioactive sample over a fixed time frame.

The Poisson distribution is

$$p(n) = \frac{\mu^n e^{-\mu}}{n!}, \tag{124}$$

where $\mu$ is the mean value and $n$ is the variable that takes on only integer values. Note that the Poisson distribution is fully described by a single parameter, $\mu$ the mean, as opposed to the Gaussian distribution, which is defined by a mean and standard deviation.

As you will show in Problem 7.5, the variance of the Poisson distribution is $\mu$. Therefore the uncertainty of a Poisson distributed variable, given by the RMSD, is $\sqrt{\mu}$. It is popular to refer to the measured value of a Poisson distributed variable as the observed value, $O$, and the mean value $\mu$ as the expected value, $E$. This notation is often used to define $\chi^2$ for Poisson distributed data:

$$\chi^2 = \sum_i \frac{(O_i - E_i)^2}{E_i}, \tag{125}$$

---

[3] Another important class of exceptions is experiments where one non-Gaussian distribution dominates the uncertainty. Yet another is experiments where measurements are near a physical cutoff, such as measurement of masses that are very close to zero, but which cannot have negative values.

where the denominator is determined by the fact that the uncertainty on $O_i$ is given by $\sqrt{O_i}$, which is roughly equivalent to $\sqrt{E_i}$ for any well-modeled system.

## 7.6 Maximum Likelihood Estimation for Nonlinear Models and Other Probability Density Functions

In Chapters 3 and 4 we introduced a method for maximum likelihood estimation for measurements with Gaussian uncertainties. There, we found the nice result that the uncertainties on the parameters of the model were also Gaussian and showed simple algebraic formulas for solving for the best fit parameters and their uncertainties. As it turns out, Gaussian uncertainties are the only type of uncertainties that produce such a simple likelihood function. We showed in Eq. 26 that

$$L(A) = e^{-\chi^2_{min}/2} e^{\frac{-(A-A^*)^2}{2\sigma_A^2}}. \tag{126}$$

Some people find it simpler to take the natural log of both sides of this equation to find

$$\ln(L(A)) = -\chi^2_{min}/2 - \frac{(A-A^*)^2}{2\sigma_A^2}, \tag{127}$$

which is an upside-down parabola. In Chapter 3 we showed how to solve algebraically for $A^*$, but there is another way. If we know what the likelihood curve looks like, then we can use it to find $\sigma_A$. If we define the maximum likelihood value as $L^*$ and we set $A = A^* + \sigma_A$, then Eq. 127 becomes

$$\ln(L(A^* \pm \sigma_A)) = \ln(L^*) - \frac{1}{2}. \tag{128}$$

From this we see that if we know the likelihood, we can use the maximum value to find $A^*$ and $\ln(L^*) - 1/2$ to find $\sigma_A$. This is not particularly useful information in the case of Gaussian uncertainties, where we can simply solve algebraically for $A^*$ and $\sigma_A$. It does become useful when dealing with other types of uncertainties, though. Whatever the model or the pdf of the uncertainties of each measurement is, we can still define the likelihood as the combined (multiplied) probabilities,

$$L(A, B, ...) = p_{total} \propto p_1 \times p_2 \times p_3 \times ..., \tag{129}$$

where $A, B, ...$ are the parameters of the model, and the $p_i$ are the uncertainty pdfs for each measurement. Of course there is no general simple solution, and of

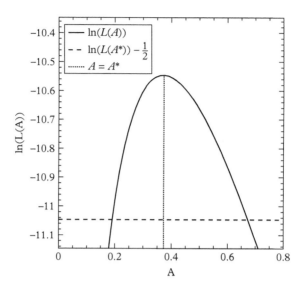

**Figure 33** *Asymmetric likelihood function resulting from four data points with log-normal and Gamma distributed uncertainties.*

course the likelihood will not be a Gaussian for non-Gaussian uncertainties and nonlinear models, but that doesn't mean that a maximum likelihood value doesn't exist or can't be found. All we need is a program that uses the information from our measurements to calculate the likelihood as a function of the parameter values, then solves numerically for the maximum value of $\ln(L)$ and the corresponding parameter values. Although the likelihood is not generally a Gaussian, it is common practice to pretend that it is and report the one-sigma uncertainty band on the parameters as given by $\ln(L^*) - 1/2$. Figure 33 shows a sketch of the uncertainty band for the likelihood of a one-parameter model with non-Guassian uncertainties. Note that the uncertainties are not necessarily symmetric, and that it is good practice to report both the lower and upper one-sigma uncertainty values separately.

## 7.7   Problems for Chapter 7

### 7.1   The mean deviation

Show that the mean deviation, $\langle (x - \langle x \rangle) \rangle$, is zero for a pdf, $p(x)$.

### 7.2   The variance identity

Show that $\mathrm{Var}[x] = \langle (x - \langle x \rangle)^2 \rangle = \langle x^2 \rangle - \langle x \rangle^2$ for a pdf, $p(x)$.

## 7.3  The *C–D* plane $\chi^2$

Show that Eq. 120 and Eq. 121 are equivalent. (Hint: use Eq. 122.)

## 7.4  The $\chi^2$ matrix

If you are familiar with matrix algebra, then you may have noticed that it is possible to write $\chi^2$ as

$$\chi^2(A, B) = \begin{bmatrix} \chi_A & \chi_B \end{bmatrix} \cdot \begin{bmatrix} 1 & \rho \\ \rho & 1 \end{bmatrix} \cdot \begin{bmatrix} \chi_A \\ \chi_B \end{bmatrix}, \tag{130}$$

where $\chi_A = (A - A^*)/\sigma_A$ and $\chi_B = (B - B^*)/\sigma_B$. When we rewrote the likelihood in terms of $C$ and $D$ in Eq. 121, all we really did was rewrite the above matrix in terms of its corresponding diagonal matrix. Use the same method to find $\chi_C$ and $\chi_D$.

## 7.5  The Poisson distribution

Show that the mean and variance of the Poisson distribution are both the mean, $\mu$.

# Appendices

The following appendices contain scripts needed to perform the one- and two-parameter linear chi-squared analyses described in this book. Newer versions of these scripts and versions for other programming languages may be available on this book's website.

## Appendix A MATLAB: One-Parameter Linear Chi-squared Script

```
% AUTHORS:  Carey Witkov and Keith Zengel
% COPYRIGHT: This script is copyright (2018) Carey Witkov and
% Keith Zengel.
% LICENSE:  Creative Commons Attribution-NonCommercial 4.0
% International
% (CC BY-NC 4.0)  https://creativecommons.org/licenses/by-nc/4.0/
% DISCLAIMER: THE SOFTWARE IS PROVIDED "AS IS", WITHOUT WARRANTY
% OF ANY KIND,
% EXPRESS OR IMPLIED, INCLUDING BUT NOT LIMITED TO THE
% WARRANTIES OF
% MERCHANTABILITY, FITNESS FOR A PARTICULAR PURPOSE AND
% NONINFRINGEMENT.
% IN NO EVENT SHALL THE AUTHORS OR COPYRIGHT HOLDERS BE LIABLE
% FOR ANY CLAIM,
% DAMAGES OR OTHER LIABILITY, WHETHER IN AN ACTION OF CONTRACT,
% TORT OR
% OTHERWISE, ARISING FROM, OUT OF OR IN CONNECTION WITH THE
% SOFTWARE OR THE
% USE OR OTHER DEALINGS IN THE SOFTWARE.


% Chi-squared curve fitting to a 1-parameter (y=Ax) linear model

clear all;
close all;

% x,y data arrays and y-error array
% replace sample data with your own data
% x is an array of mean values for the independent variable
% y is an array of mean values for the dependent variable
% yerr is an array of standard errors (i.e. SD/(sqrt of N)) for y
x=[0.10 0.60 1.10 1.60 2.10 2.60 3.10 3.60 4.10 4.60 5.10 5.60 ...
   6.10 6.60 7.10 7.60 8.10 8.60 9.10 9.60];
y=[34.1329 98.7892 121.0725 180.3328 236.3683 260.5684 ...
   320.9553 380.3028 407.3759 453.7503 506.9685 576.329 ...
```

```
        602.0845 699.0915 771.2271 796.6707 787.8436 877.0763 ...
        915.3649 1000.7312];
yerr=[5.8423 9.9393 11.0033 13.4288 15.3743 16.1421 17.9152 ...
        19.5014 20.1836 21.3014 22.516 24.0194 24.5374 26.4403 ...
        27.771 28.2254 28.0686 29.6155 30.255 31.6343];

%%%%%%%%%%%%%%%%%%%%%%%%%%%%%%%%%%%%%%%%%%%%%%%%%%%%%%%%%%%%%%%%
%%%%%%%%%%%%% DO NOT EDIT BELOW THIS LINE %%%%%%%%%%%%%%%%%%%%%
%%%%%%%%%%%%%%%%%%%%%%%%%%%%%%%%%%%%%%%%%%%%%%%%%%%%%%%%%%%%%%%%

% calculate sums needed to obtain chi-square
s_yy=sum(y.^2./yerr.^2);
s_xx=sum(x.^2./yerr.^2);
s_xy=sum((y.*x)./yerr.^2);

% by completing the square, we rewrite chi-squared as
% sum((y_i - A x_i)^2/sigma_i^2 = (A-A^*)^2/sigma_A^2 +
% \chi^2_{min}
A_best = s_xy/s_xx
sigma_A = 1/sqrt(s_xx)
minchi2 = s_yy - s_xy^2/s_xx

%define chi-squared plot interval as A^* +- 2*sigma_A
twosigma = [A_best-2*sigma_A A_best+2*sigma_A];

%plot chi-squared parabola and minchi+1
figure(1)
fplot( @(A) (A - A_best).^2/sigma_A.^2 + minchi2, twosigma);
xlabel('A','Fontsize',16)
ylabel('\chi^2','Fontsize',16)
title('\chi^2 vs. slope and \chi^2_{min}+1','Fontsize',16)
hold on;
fplot(minchi2+1, twosigma);

%define fit plot interval around x_min and x_max
delta_x  = range(x);
xmintoxmax = [min(x)-0.1*delta_x max(x)+0.1*delta_x];

%plot data and line of best fit
figure(2)
errorbar(x, y, yerr,'x')
xlabel('x','Fontsize',16)
ylabel('y','Fontsize',16)
title('y vs x data and line-of-best-fit','Fontsize',16)
hold on;
fplot(@(x) A_best.*x, xmintoxmax)
hold off;
```

## Appendix B MATLAB: Two-Parameter Linear Chi-squared Script

```
% AUTHORS:  Carey Witkov and Keith Zengel
% COPYRIGHT: This script is copyright (2018) Carey Witkov and
% Keith Zengel.
% LICENSE:  Creative Commons Attribution-NonCommercial 4.0
% International
% (CC BY-NC 4.0)  https://creativecommons.org/licenses/by-nc/4.0/
% DISCLAIMER: THE SOFTWARE IS PROVIDED "AS IS", WITHOUT WARRANTY
% OF ANY KIND,
% EXPRESS OR IMPLIED, INCLUDING BUT NOT LIMITED TO THE
% WARRANTIES OF
% MERCHANTABILITY, FITNESS FOR A PARTICULAR PURPOSE AND
% NONINFRINGEMENT.
% IN NO EVENT SHALL THE AUTHORS OR COPYRIGHT HOLDERS BE LIABLE
% FOR ANY CLAIM,
% DAMAGES OR OTHER LIABILITY, WHETHER IN AN ACTION OF CONTRACT,
% TORT OR
% OTHERWISE, ARISING FROM, OUT OF OR IN CONNECTION WITH THE
% SOFTWARE OR THE
% USE OR OTHER DEALINGS IN THE SOFTWARE.

% Chi-squared curve fitting to a 2-parameter (y=Ax+b) linear
% model

clear all;
close all;

% x,y data arrays and y-error array
% replace sample data with your own data
% x is an array of mean values for the independent variable
% y is an array of mean values for the dependent variable
% yerr is an array of standard errors (i.e. SD/(sqrt of N)) for y
x=[0.10 0.60 1.10 1.60 2.10 2.60 3.10 3.60 4.10 4.60 5.10 5.60 ...
  6.10 6.60 7.10 7.60 8.10 8.60 9.10 9.60]   ;
y=[34.1329 98.7892 121.0725 180.3328 236.3683 260.5684 ...
  320.9553 380.3028 407.3759 453.7503 506.9685 576.329 ...
  602.0845 699.0915 771.2271 796.6707 787.8436 877.0763 ...
  915.3649 1000.7312];
yerr=[5.8423 9.9393 11.0033 13.4288 15.3743 16.1421 17.9152 ...
  19.5014 20.1836 21.3014 22.516 24.0194 24.5374 26.4403 ...
  27.771 28.2254 28.0686 29.6155 30.255 31.6343];
```

```
%%%%%%%%%%%%%%%%%%%%%%%%%%%%%%%%%%%%%%%%%%%%%%%%%%%%%%%%%%%%%%%
%%%%%%%%%%%%% DO NOT EDIT BELOW THIS LINE %%%%%%%%%%%%%%%%%%%%
%%%%%%%%%%%%%%%%%%%%%%%%%%%%%%%%%%%%%%%%%%%%%%%%%%%%%%%%%%%%%%%

% calculate sums needed to obtain chi-square
s_yy=sum(y.^2./yerr.^2);
s_xx=sum(x.^2./yerr.^2);
s_xy=sum((y.*x)./yerr.^2);
s_y =sum(y./yerr.^2);
s_x =sum(x./yerr.^2);
s_0 =sum(1./yerr.^2);

% by completing the square, we rewrite chi-squared as
% sum((y_i - A x_i - B)^2/sigma_i^2
% = (A-A^*)^2/sigma_A^2
% + (B-B^*)^2/sigma_B^2
% + 2*rho*(A-A^*)(B-B^*)/sigma_A*Sigma_A
% + \chi^2_{min}
A_best = (s_0*s_xy - s_x*s_y)/(s_0*s_xx - s_x^2)
sigma_A = 1/sqrt(s_xx);
B_best = (s_y*s_xx - s_x*s_xy)/(s_0*s_xx - s_x^2)
sigma_B = 1/sqrt(s_0);
rho = s_x/sqrt(s_xx*s_0);
minchi2 = (s_0*s_xy^2 - 2*s_x*s_y*s_xy + s_y^2*s_xx)/(s_x^2 -
 s_0*s_xx) + s_yy

%define chi-squared plot interval
A_interval = sqrt(6.17*sigma_A^2/(1-rho^2));
B_interval = sqrt(6.17*sigma_B^2/(1-rho^2));
AB_interval = [B_best-1.1*B_interval B_best+1.1*B_interval
 A_best-1.1*A_interval A_best+1.1*A_interval];

% Contour plot
figure(1);
paraboloid = @(B,A) (A-A_best).^2/sigma_A.^2 + (B-B_best).^2./
 sigma_B.^2 + 2*rho*(A-A_best).*(B-B_best)./(sigma_A.*sigma_B);
fcontour(paraboloid, AB_interval,'LevelList',[2.3 6.17],
 'MeshDensity',200);
xlabel('B (intercept)','Fontsize',16);
ylabel('A (slope)','Fontsize',16);
title('68% and 95% contours of \chi^2 in the AB plane',
 'Fontsize',16);
hold on;
plot(B_best,A_best, 'x');
hold off;
```

```
%define fit plot interval around x_min and x_max
delta_x  = range(x);
xmintoxmax = [min(x)-0.1*delta_x max(x)+0.1*delta_x];

%plot data and line of best fit
figure(2); clf;
errorbar(x, y, yerr,'x');
xlabel('x','Fontsize',16);
ylabel('y','Fontsize',16);
title('y vs x data and line-of-best-fit','Fontsize',16);
hold on;
fplot(@(x) A_best.*x + B_best, xmintoxmax);
hold off;
```

## Appendix C Python: One-Parameter Linear Chi-squared Script

```
# py1.py version 0.2
# Chi-squared curve fitting to a 1-parameter (y=Ax) linear model
# Three arrays are needed:
# x is an array of mean values for the independent variable
# y is an array of mean values for the dependent variable
# yerr is an array of standard errors (i.e. SD/(sqrt of N))
# for y
# Note that this script only handles errors on the dependent
# (y) variable.
# SOFTWARE DEPENDENCIES:  Python 3, Numpy, Matplotlib
# -----------------------------------------------------------
# AUTHORS:  Carey Witkov and Keith Zengel
# COPYRIGHT: This script is copyright (2018) Carey Witkov and
# Keith Zengel.
# LICENSE:  Creative Commons Attribution-NonCommercial 4.0
# International
# (CC BY-NC 4.0)  https://creativecommons.org/licenses/by-nc/4.0/
# DISCLAIMER: THE SOFTWARE IS PROVIDED "AS IS", WITHOUT WARRANTY
# OF ANY KIND,
# EXPRESS OR IMPLIED, INCLUDING BUT NOT LIMITED TO THE
# WARRANTIES OF
# MERCHANTABILITY, FITNESS FOR A PARTICULAR PURPOSE AND
# NONINFRINGEMENT.
# IN NO EVENT SHALL THE AUTHORS OR COPYRIGHT HOLDERS BE LIABLE
# FOR ANY CLAIM,
# DAMAGES OR OTHER LIABILITY, WHETHER IN AN ACTION OF CONTRACT,
# TORT OR
# OTHERWISE, ARISING FROM, OUT OF OR IN CONNECTION WITH THE
# SOFTWARE OR THE
# USE OR OTHER DEALINGS IN THE SOFTWARE.
```

```python
import matplotlib.pyplot as plt
from numpy import *
import numpy as np

# data input
x=array([0.1, 0.6, 1.1, 1.6, 2.1, 2.6, 3.1, 3.6, 4.1, 4.6, 5.1, \
    5.6, 6.1, 6.6, 7.1, 7.6, 8.1, 8.6, 9.1, 9.6])
y=array([34.1329, 98.7892, 121.0725, 180.3328, 236.3683, \
    260.5684, 320.9553, 380.3028, 407.3759, 453.7503, 506.9685, \
    576.9329,602.0845, 699.0915, 771.2271, 796.6707, 787.8436, \
    877.0763, 915.3649, 1000.7312])
yerr=array([5.8423, 9.9393, 11.0033, 13.4288, 15.3743, 16.1421, \
    17.9152, 19.5014, 20.1836, 21.3014, 22.516, 24.0194, \
    24.5374, 26.4403, 27.771, 28.2254, 28.0686, 29.6155, \
    30.255, 31.6343])

# calculate sums needed to obtain chi-square
s_yy=sum(y**2/yerr**2)
s_xx=sum(x**2/yerr**2)
s_xy=sum((y*x)/yerr**2)

# by completing the square, we rewrite chi-squared as
# sum((y_i - A x_i)^2/sigma_i^2 = (A-A^*)^2/sigma_A^2 +
# \chi^2_{min}
A_best = s_xy/s_xx
sigma_A = 1/sqrt(s_xx)
minchi2 = s_yy - s_xy**2/s_xx

# define chi-squared plot interval as A^* +- 2*sigma_A
twosigma = array([A_best-2*sigma_A, A_best+2*sigma_A])

# create parameter range for slope
a = np.linspace(twosigma[0], twosigma[1],1000)

# calculate chi-square over parameter grid
chi2=s_yy + (a**2)*s_xx - 2*a*s_xy;

# plot data with errorbars
plt.figure(1)
plt.plot(x,A_best*x)
plt.errorbar(x,y,yerr,linestyle='None',fmt='.k')
plt.xlabel('x', fontsize=16)
plt.ylabel('y', fontsize=16)
plt.grid(True)
plt.title("y vs x data with y-error bars")
```

```
# display chi-square vs. slope
plt.figure(2)
plt.plot(a,chi2,'o')
plt.axhline(y=minchi2+1, color='r')
plt.xlabel('slope',fontsize=16)
plt.ylabel('chisq',fontsize=16)
plt.grid(True)
plt.title("Chi-square as a function of slope \n %4d points \
  chisq min \
  =%6.2f best slope =%7.2f " %(x.size,minchi2,A_best))
```

## Appendix D Python: Two-Parameter Linear Chi-squared Script

```
# py2.py version 0.2
# Chi-squared curve fitting to a 2-parameter (y=Ax+b) linear model
# Three arrays are needed:
# x is an array of mean values for the independent variable
# y is an array of mean values for the dependent variable
# yerr is an array of standard errors (i.e. SD/(sqrt of N)) for y
# Note that this script only handles errors on the dependent
# (y) variable.
# SOFTWARE DEPENDENCIES:  Python 3, Numpy, Matplotlib
# ---------------------------------------------------------------
# AUTHORS:  Carey Witkov and Keith Zengel
# COPYRIGHT: This script is copyright (2018) Carey Witkov and
# Keith Zengel.
# LICENSE:  Creative Commons Attribution-NonCommercial 4.0
# International
# (CC BY-NC 4.0)  https://creativecommons.org/licenses/by-nc/4.0/
# DISCLAIMER: THE SOFTWARE IS PROVIDED "AS IS", WITHOUT WARRANTY
# OF ANY KIND,
# EXPRESS OR IMPLIED, INCLUDING BUT NOT LIMITED TO THE
# WARRANTIES OF
# MERCHANTABILITY, FITNESS FOR A PARTICULAR PURPOSE AND
# NONINFRINGEMENT.
# IN NO EVENT SHALL THE AUTHORS OR COPYRIGHT HOLDERS BE LIABLE
# FOR ANY CLAIM,
# DAMAGES OR OTHER LIABILITY, WHETHER IN AN ACTION OF CONTRACT,
# TORT OR
# OTHERWISE, ARISING FROM, OUT OF OR IN CONNECTION WITH THE
# SOFTWARE OR THE
# USE OR OTHER DEALINGS IN THE SOFTWARE.

import matplotlib.pyplot as plt
from numpy import *
import numpy as np
```

```
# data input
x=array([0.1, 0.6, 1.1, 1.6, 2.1, 2.6, 3.1, 3.6, 4.1, 4.6, 5.1, \
  5.6, 6.1, 6.6, 7.1, 7.6, 8.1, 8.6, 9.1, 9.6])
y=array([34.1329, 98.7892, 121.0725, 180.3328, 236.3683, \
  260.5684, 320.9553, 380.3028, 407.3759, 453.7503, 506.9685, \
  576.9329,602.0845, 699.0915, 771.2271, 796.6707, 787.8436, \
  877.0763, 915.3649, 1000.7312])
yerr=array([5.8423, 9.9393, 11.0033, 13.4288, 15.3743, 16.1421, \
  17.9152, 19.5014, 20.1836, 21.3014, 22.516, 24.0194, \
  24.5374, 26.4403, 27.771, 28.2254, 28.0686, 29.6155, 30.255, \
  31.6343])

# calculate sums needed to obtain chi-square
s_yy=sum(y**2/yerr**2)
s_xx=sum(x**2/yerr**2)
s_0=sum(1/yerr**2)
s_xy=sum((y*x)/yerr**2)
s_y=sum(y/yerr**2)
s_x=sum(x/yerr**2)

# by completing the square, we rewrite chi-squared as
# sum((y_i - A x_i - B)^2/sigma_i^2
# = (A-A^*)^2/sigma_A^2
# + (B-B^*)^2/sigma_B^2
# + 2*rho*(A-A^*)(B-B^*)/sigma_A*Sigma_A
# + \chi^2_{min}
A_best = (s_0*s_xy - s_x*s_y)/(s_0*s_xx - s_x**2)
sigma_A = 1/sqrt(s_xx);
B_best = (s_y*s_xx - s_x*s_xy)/(s_0*s_xx - s_x**2)
sigma_B = 1/sqrt(s_0);
rho = s_x/sqrt(s_xx*s_0);
minchi2 = (s_0*s_xy**2 - 2*s_x*s_y*s_xy + s_y**2*s_xx)/(s_x**2 - \
  s_0*s_xx) + s_yy

# create parameter grid
A_interval = 1.1*(sqrt(6.17*sigma_A**2/(1-rho**2)));
B_interval = 1.1*(sqrt(6.17*sigma_B**2/(1-rho**2)));

# create parameter grid
a = np.linspace(A_best-A_interval, A_best+A_interval)
b = np.linspace(B_best-B_interval, B_best+B_interval)
A,B = np.meshgrid(a,b)

# calculate chi-square over parameter grid
# chi2=(S1) + (A**2)*(S2) + (B**2)*(S3) - 2*A*S4 - 2*B*S5 +
# 2*A*B*S6
```

```
chi2=s_yy + (A**2)*s_xx + (B**2)*s_0 - 2*A*s_xy - 2*B*s_y + \
  2*A*B*s_x;

# plot data with errorbars
plt.figure(1)
plt.plot(x,A_best*x)
plt.errorbar(x,y,yerr,linestyle='None',fmt='.k')
plt.xlabel('x', fontsize=16)
plt.ylabel('y', fontsize=16)
plt.grid(True)
plt.title("y vs x data with y-error bars")

#plot chi-square in A-b parameter plane with 68% and 95% contours
plt.figure(2)
levels=[minchi2,minchi2+2.3,minchi2+6]
Z=plt.contour(B,A,chi2,levels)
plt.clabel(Z,inline=1, fontsize=10)
plt.plot(B_best,A_best,'+')
plt.xlabel('B (intercept)',fontsize=16)
plt.ylabel('A (slope)',fontsize=16)
plt.title('Chi-square 68% and 95% contours in A-B plane')
```

## Appendix E Solutions

### E.1   Solutions for Chapter 2

#### 2.1   *Average or calculate first?*

Right away we know that averaging first gives us $f(\langle y \rangle)$. What we want to find is $\langle f(y) \rangle$. First, we Taylor expand $f(y)$ around $f(\langle y \rangle)$:

$$f(y) \approx f(\langle y \rangle) + f'(\langle y \rangle)(y - \langle y \rangle) + \frac{1}{2}f''(\langle y \rangle)(y - \langle y \rangle)^2. \tag{E1}$$

Next, we find the average:

$$\langle f(y) \rangle \approx \langle f(\langle y \rangle) \rangle + f'(\langle y \rangle)\langle (y - \langle y \rangle) \rangle + \frac{1}{2}f''(\langle y \rangle)\langle (y - \langle y \rangle)^2 \rangle. \tag{E2}$$

The middle term is zero because the mean deviation is zero. The last term is $f''(\langle y \rangle)$ times the mean squared deviation, so

$$\langle f(y) \rangle - f(\langle y \rangle) = \frac{1}{2}f''(\langle y \rangle)(\text{RMSD})^2. \tag{E3}$$

This term is sometimes called the *bias*.

In the case of $\sqrt{2d/g}$, the bias is $-\sqrt{1/8g\langle d \rangle^3}$. Therefore the calculate-then-average value will always be lower than the average-then-calculate value.

## 2.2   Standard error derivation

The mean of a set of measured values of $y$ is

$$\langle y \rangle = \frac{1}{N} \sum_{i=1}^{N} y_i. \tag{E4}$$

If we assume that the uncertainty associated with each measurement is its deviation from the mean, then

$$\sigma_f^2 = \sum_{j=1}^{N} \left| \frac{\partial f}{\partial y_j} \right|^2 \sigma_{y_j}^2$$

$$= \sum_{j=1}^{N} \left| \frac{\partial}{\partial y_j} \frac{1}{N} \sum_{i=1}^{N} y_i \right|^2 (y_j - \langle y \rangle)^2$$

$$= \sum_{j=1}^{N} \left| \frac{1}{N} \right|^2 (y_j - \langle y \rangle)^2 \tag{E5}$$

$$= \frac{(\text{RMSD})^2}{N}.$$

Taking the square root of both sides gives the expected result of

$$\text{SE} = \frac{\text{RMSD}}{\sqrt{N}}. \tag{E6}$$

## 2.3   Correlated uncertainties

Taylor expanding a function of several variables gives

$$f(x, y, z, ...) \approx f(\langle x \rangle, \langle y \rangle, \langle z \rangle, ...) + \frac{\partial f}{\partial x}(x - \langle x \rangle) + \frac{\partial f}{\partial y}(y - \langle y \rangle)$$

$$+ \frac{\partial f}{\partial z}(z - \langle z \rangle) + ..., \tag{E7}$$

where all of the partial derivatives are evaluated at the mean values. Then to second order the mean square deviation is

$$\langle (f(x, y, z, ...) - f(\langle x \rangle, \langle y \rangle, \langle z \rangle, ...))^2 \rangle = \left| \frac{\partial f}{\partial x} \right|^2 \langle (x - \langle x \rangle)^2 \rangle + \left| \frac{\partial f}{\partial y} \right|^2 \langle (y - \langle y \rangle)^2 \rangle +$$

$$\left| \frac{\partial f}{\partial z} \right|^2 \langle (z - \langle z \rangle)^2 \rangle + 2 \frac{\partial f}{\partial x} \frac{\partial f}{\partial y} \langle (x - \langle x \rangle)(y - \langle y \rangle) \rangle + 2 \frac{\partial f}{\partial x} \frac{\partial f}{\partial z} \langle (x - \langle x \rangle)(z - \langle z \rangle) \rangle + \tag{E8}$$

$$2 \frac{\partial f}{\partial y} \frac{\partial f}{\partial z} \langle (y - \langle y \rangle)(z - \langle z \rangle) \rangle + ...$$

So

$$
\sigma_f^2 = \left|\frac{\partial f}{\partial x}\right|^2 \sigma_x^2 + \left|\frac{\partial f}{\partial y}\right|^2 \sigma_y^2 + \left|\frac{\partial f}{\partial z}\right|^2 \sigma_z^2 + 2\frac{\partial f}{\partial x}\frac{\partial f}{\partial y}\sigma_{xy} + 2\frac{\partial f}{\partial x}\frac{\partial f}{\partial z}\sigma_{xz} + 2\frac{\partial f}{\partial y}\frac{\partial f}{\partial z}\sigma_{yz}. \quad \text{(E9)}
$$

## 2.4 Divide and conquer

Using the result from Problem 2.3, we find

$$
\begin{aligned}
\sigma_f^2 &= \left|\frac{\partial f}{\partial x}\right|^2 \sigma_x^2 + \left|\frac{\partial f}{\partial y}\right|^2 \sigma_y^2 + 2\frac{\partial f}{\partial x}\frac{\partial f}{\partial y}\sigma_{xy} \\
&= \left|\frac{1}{y}\right|^2 \sigma_x^2 + \left|\frac{-x}{y^2}\right|^2 \sigma_y^2 + 2\left(\frac{1}{y}\right)\left(\frac{-x}{y^2}\right)\sigma_{xy} \quad \text{(E10)} \\
&= \frac{1}{y^2}\sigma_x^2 + \frac{x^2}{y^4}\sigma_y^2 - 2\frac{x}{y^3}\sigma_{xy},
\end{aligned}
$$

where $x$ and $y$ are evaluated at $\langle x\rangle$ and $\langle y\rangle$. To see how this uncertainty compares with the individual uncertainties, we can look at the fractional uncertainty, $\sigma_f/f$:

$$
\frac{\sigma_f^2}{f^2} = \frac{1}{x^2}\sigma_x^2 + \frac{1}{y^2}\sigma_y^2 - 2\frac{1}{xy}\sigma_{xy}. \quad \text{(E11)}
$$

From this we can see that if $\sigma_{xy}$ is positive, meaning that $x$ and $y$ are positively correlated, then the uncertainty on the ratio can be smaller than the uncertainty on $x$ or $y$. This means that it is beneficial to report the ratio of correlated variables, since the total uncertainty will be reduced.

## 2.5 Be fruitful and multiply

As in Problem 2.4, we begin with

$$
\begin{aligned}
\sigma_f^2 &= \left|\frac{\partial f}{\partial x}\right|^2 \sigma_x^2 + \left|\frac{\partial f}{\partial y}\right|^2 \sigma_y^2 + 2\frac{\partial f}{\partial x}\frac{\partial f}{\partial y}\sigma_{xy} \\
&= |y|^2 \sigma_x^2 + |x|^2 \sigma_y^2 + 2xy\sigma_{xy},
\end{aligned} \quad \text{(E12)}
$$

where $x$ and $y$ are evaluated at $\langle x\rangle$ and $\langle y\rangle$. Then the fractional uncertainty is

$$
\frac{\sigma_f^2}{f^2} = \frac{1}{x^2}\sigma_x^2 + \frac{1}{y^2}\sigma_y^2 + \frac{2}{xy}\sigma_{xy}. \quad \text{(E13)}
$$

From this we can see that if $\sigma_{xy}$ is negative, meaning that $x$ and $y$ are negatively correlated (anti-correlated), then the uncertainty on the ratio can be smaller than the uncertainty on $x$ or $y$. This means that it is beneficial to report the product of anti-correlated variables, since the total uncertainty will be reduced.

## 2.6   Negligible uncertainties

First, we propagate the uncertainty for a sum of two variables,

$$\sigma_f^2 = \left|\frac{\partial f}{\partial x}\right|^2 \sigma_x^2 + \left|\frac{\partial f}{\partial y}\right|^2 \sigma_y^2$$

$$= \sigma_x^2 + \sigma_y^2.$$

(E14)

This result is sometimes referred to as *adding in quadrature*. For $\sigma_x = 5\sigma_y$ and $\sigma_x = 10\sigma_y$, we have

$$\sigma_f = \sqrt{26}\sigma_y \approx \sigma_x$$

(E15)

and

$$\sigma_f = \sqrt{101}\sigma_y \approx \sigma_x.$$

(E16)

## E.2   Solutions for Chapter 3

### 3.1   Least absolute sum minimization

To solve this problem using algebra, we can rewrite the sum inside the absolute value as

$$\left|\sum_i (y_i - Ax_i)\right| = \left|\sum_i y_i - A\sum_i x_i\right|.$$

(E17)

Because of the absolute value, the minimum value that this function can take on is zero. Therefore,

$$\sum_i y_i - A\sum_i x_i = 0, \text{ or}$$

$$A = \frac{\sum_i y_i}{\sum_i x_i}.$$

(E18)

To solve this problem using calculus, we'll need to know that the derivative of an absolute value is

$$\frac{d|x|}{dx} = \frac{x}{|x|}.$$

(E19)

Note that the slope is $\pm 1$. In our case we want the minimum with respect to $A$, so

$$\frac{d}{dA}\left|\sum_i (y_i - Ax_i)\right| = \frac{\sum_i (y_i - Ax_i)}{\left|\sum_i (y_i - Ax_i)\right|} = 0.$$

(E20)

So we again find that

$$A = \frac{\sum_i y_i}{\sum_i x_i}.$$

(E21)

## 3.2   Least squares minimization

To solve this problem using algebra, we can expand the numerator to find

$$C = \sum_i (y_i - Ax_i)^2$$

$$= \sum_i y_i^2 + 2A \sum_i y_i x_i + A^2 \sum_i x_i^2.$$

(E22)

Then we can rename the sums $S_{yy}$, $S_{xy}$, and $S_{xx}$, to save a little writing:

$$S_{yy} = \sum_i y_i^2,$$

$$S_{xy} = \sum_i y_i x_i, \text{ and}$$

$$S_{xx} = \sum_i x_i^2,$$

(E23)

so

$$C(A) = S_{yy} + 2AS_{xy} + A^2 S_{xx}.$$

(E24)

We want to complete the square so that our equation is of the form

$$C(A) = \frac{(A - A^*)^2}{\sigma_A^2} + C_{min},$$

(E25)

which we can rewrite as

$$C(A) = \frac{1}{\sigma_A^2} A^2 + 2A \frac{A^*}{\sigma_A^2} + \left[ \frac{(A^*)^2}{\sigma_A^2} + C_{min} \right].$$

(E26)

Next we can set the coefficients in Eq. E24 equal to those in Eq. E26, to find

$$\frac{1}{\sigma_A^2} = S_{xx},$$

$$\frac{A^*}{\sigma_A^2} = S_{xy}, \text{ and}$$

$$\left[ \frac{(A^*)^2}{\sigma_A^2} + C_{min} \right] = S_{yy}.$$

(E27)

We can use these three equations to solve for $\sigma_A$, $A^*$, and $C_{min}$:

$$\sigma_A = \frac{1}{\sqrt{S_{xx}}},$$

$$A^* = \frac{S_{xy}}{S_{xx}}, \text{ and} \tag{E28}$$

$$C_{min} = S_{yy} - \frac{S_{xy}^2}{S_{xx}^2}.$$

Notice that this cost function has a minimum value $C_{min}$ at $A^*$.

To solve this problem with calculus, we can take a derivative with respect to $A$ and set the result equal to zero,

$$\frac{dC(A)}{dA}\Big|_{A_{min}} = 2S_{xy} + 2A_{min}S_{xx} = 0 , \tag{E29}$$

or

$$A_{min} = \frac{S_{xy}}{S_{yy}}. \tag{E30}$$

Note that $A_{min}$ is the same as $A^*$ previously.

## 3.3   $\chi^2$ *minimization*

The solution is the same as for Problem 3.2, except with $S_{xx}$, $S_{xy}$, and $S_{yy}$ defined as

$$S_{yy} = \sum_i \frac{y_i^2}{\sigma_i^2},$$

$$S_{xy} = \sum_i \frac{y_i x_i}{\sigma_i^2}, \text{ and} \tag{E31}$$

$$S_{xx} = \sum_i \frac{x_i^2}{\sigma_i^2}.$$

Note that the two results are identical if all of the uncertainties are one unit of measurement ($\sigma_i = 1$).

## 3.4   *The RMSD of* $\chi^2$

To save some writing, let's consider the mean squared deviation of $\chi^2$, which is really the variance of $\chi^2$:

$$\begin{aligned} \text{Var}(\chi^2) &= \langle(\chi^2 - \langle\chi^2\rangle)\rangle \\ &= \langle(\chi^2)^2\rangle - 2\langle\chi^2\rangle^2 + \langle\chi^2\rangle^2 \\ &= \langle(\chi^2)^2\rangle - \langle\chi^2\rangle^2. \end{aligned} \tag{E32}$$

The real difficulty here is in calculating $\langle(\chi^2)^2\rangle$. To calculate $(\chi^2)^2$, we need to multiply two sums,

$$(\chi^2)^2 = \left[\sum_i^N \frac{(y_i - Ax_i)^2}{\sigma_i^2}\right]\left[\sum_j^N \frac{(y_j - Ax_j)^2}{\sigma_j^2}\right]. \tag{E33}$$

Note that the indices of the two sums are different. We can break this up into the $N$ terms where $i = j$ and the $N^2 - N$ terms where $i \neq j$:

$$(\chi^2)^2 = \left[\sum_{i=j} \frac{(y_i - Ax_i)^2(y_j - Ax_j)^2}{\sigma_i^2 \sigma_j^2}\right] + \left[\sum_{i \neq j} \frac{(y_i - Ax_i)^2(y_j - Ax_j)^2}{\sigma_i^2 \sigma_j^2}\right]$$

$$(\chi^2)^2 = \left[\sum_{i=1}^{N} \frac{(y_i - Ax_i)^4}{\sigma_i^4}\right] + \left[\sum_{i \neq j} \frac{(y_i - Ax_i)^2(y_j - Ax_j)^2}{\sigma_i^2 \sigma_j^2}\right].$$

(E34)

Next we can take the expected value of both sides to find

$$\langle(\chi^2)^2\rangle = \left[\sum_{i=1}^{N} \frac{\langle(y_i - Ax_i)^4\rangle}{\sigma_i^4}\right] + \left[\sum_{i \neq j} \frac{\langle(y_i - Ax_i)^2(y_j - Ax_j)^2\rangle}{\sigma_i^2 \sigma_j^2}\right], \qquad (E35)$$

which we can simplify using the identities $\langle(x - \mu)^4\rangle = 3\sigma^4$ and $\langle x \cdot y\rangle = \langle x\rangle\langle y\rangle$:

$$\langle(\chi^2)^2\rangle = \left[\sum_{i=1}^{N} \frac{3\sigma_i^4}{\sigma_i^4}\right] + \left[\sum_{i \neq j} \frac{\sigma_i^2 \sigma_j^2}{\sigma_i^2 \sigma_j^2}\right]$$

$$= [3N] + \left[N^2 - N\right].$$

(E36)

We can plug this result back into Eq. E32 to find

$$\text{Var}(\chi^2) = \langle(\chi^2)^2\rangle - \langle\chi^2\rangle^2$$

$$= 3N + N^2 - N - N^2 \qquad (E37)$$

$$= 2N.$$

Finally, we can take the square root both of sides to find the desired result,

$$\text{RMSD}(\chi^2) = \sqrt{2N}. \qquad (E38)$$

### 3.5 *More is better*

The benefit of taking more measurements is that the relative (fractional) uncertainty of chi-squared decreases with more measurements. We can define the relative uncertainty as the ratio of the RMSD to the mean value of chi-squared,

$$\sigma_{\text{rel}} = \frac{\text{RMSD}(\chi^2)}{\langle\chi^2\rangle} = \frac{\sqrt{2N}}{N} = \sqrt{\frac{2}{N}}. \qquad (E39)$$

From this we see that the relative uncertainty decreases as $\sqrt{2/N}$, which means that more measurements should lead to smaller uncertainties. This is a more rigorous confirmation of the intuitive idea that your result is more believable with more measurements.

## E.3 Solutions for Chapter 4

### 4.1 Completing the square

To solve for the desired parameters in terms of the $S$'s, we should start by expanding all the terms:

$$
\chi^2 = \frac{(A - A^*)^2}{\sigma_A^2} + \frac{(B - B^*)^2}{\sigma_B^2} + 2\rho\frac{(A - A^*)(B - B^*)}{\sigma_A\sigma_B} + \chi_{\min}^2
$$

$$
= A^2\frac{1}{\sigma_A^2} + -2A\frac{A^*}{\sigma_A^2} + \frac{(A^*)^2}{\sigma_A^2} + B^2\frac{1}{\sigma_B^2} + -2B\frac{B^*}{\sigma_B^2} + \frac{(B^*)^2}{\sigma_B^2} \tag{E40}
$$

$$
+ 2AB\frac{\rho}{\sigma_A\sigma_B} - 2A\frac{\rho B^*}{\sigma_A\sigma_B} - 2B\frac{\rho A^*}{\sigma_A\sigma_B} + 2\frac{\rho A^* B^*}{\sigma_A\sigma_B} + \chi_{\min}^2.
$$

Next, we set the coefficients equal with those from

$$
\chi^2 = [S_{yy}] - 2A[S_{xy}] - 2B[S_y] + -2AB[S_x] + A^2[S_{xx}] + B^2[S_0], \tag{E41}
$$

to find

$$
\frac{(A^*)^2}{\sigma_A^2} + \frac{(B^*)^2}{\sigma_B^2} + 2\frac{\rho A^* B^*}{\sigma_A\sigma_B} + \chi_{\min}^2 = S_{yy},
$$

$$
\frac{A^*}{\sigma_A^2} + \frac{\rho B^*}{\sigma_A\sigma_B} = S_{xy},
$$

$$
\frac{B^*}{\sigma_B^2} + \frac{\rho A^*}{\sigma_A\sigma_B} = S_y, \tag{E42}
$$

$$
\frac{\rho}{\sigma_A\sigma_B} = S_x,
$$

$$
\frac{1}{\sigma_A^2} = S_{xx}, \text{ and}
$$

$$
\frac{1}{\sigma_B^2} = S_0.
$$

Rearranging these equations gives

$$
\sigma_A = \frac{1}{\sqrt{S_{xx}}},
$$

$$
\sigma_B = \frac{1}{\sqrt{S_0}},
$$

$$
\rho = \frac{S_x}{\sqrt{S_{xx}S_0}},
$$

$$
A^* = \frac{(S_0 S_{xy} - S_x S_y)}{(S_0 S_{xx} - S_x^2)}, \tag{E43}
$$

$$
B^* = \frac{(S_y S_{xx} - S_x S_{xy})}{(S_0 S_{xx} - S_x^2)}, \text{ and}
$$

$$
\chi_{\min}^2 = S_{yy} + \frac{(S_0 S_{xy}^2 - 2S_x S_y S_{xy} + S_{xx} S_y^2)}{(S_x^2 - S_0 S_{xx})}.
$$

## 4.2   Reference parameter values

If you know the true value of the parameters you are trying to estimate, you can plot them and see if they land inside our 68% or 95% contours. If they do, then you can say that your model is consistent with the data. If they do not, and if you are sure your experiment was done well, then you can reject the model.

## E.4   Solutions for Chapter 5

### 5.1   Nonlinear one-parameter models

For

$$n(t) = e^{\frac{-t}{\tau}},$$ (E44)

we can take the log of both sides to find

$$\ln(n) = -\frac{1}{\tau}t.$$ (E45)

For

$$h(r) = \ln(\frac{r}{r_0}),$$ (E46)

we can exponentiate both sides to find

$$e^h = \frac{1}{r_0}r.$$ (E47)

The last two equations,

$$v(d) = \frac{1}{\sqrt{z}}d$$ (E48)

and

$$\omega = \frac{\beta}{(1+\beta)}\Omega$$ (E49)

do not need to be modified.

## E.5   Solutions for Chapter 6

### 6.1   Nonlinear two-parameter models

For

$$E(t) = \sin(\omega t + \delta),$$ (E50)

we can use the angle-addition trigonometric identity to find

$$E = \sin(\omega t)\cos(\delta) + \cos(\omega t)\sin(\delta),$$ (E51)

which we can rearrange into the form

$$\frac{E}{\cos(\omega t)} = \cos(\delta)\tan(\omega t) + \tan(\delta).$$ (E52)

Alternatively, we can take the inverse sine of both sides to find

$$\sin^{-1}(E) = \omega t + \delta.$$ (E53)

For

$$\phi(\theta) = \frac{1}{\sqrt{\alpha + \beta \cos(\theta)}},$$

(E54)

we can square and invert to find

$$\frac{1}{\phi^2} = \alpha + \beta \cos(\theta).$$

(E55)

For

$$R(v) = Av^2 + Bv,$$

(E56)

one solution is to divide both sides by a single $v$ to find

$$\frac{R}{v} = Av + B.$$

(E57)

### 6.2 Calculating the Reynolds number

At terminal velocity, our force equation is

$$mg = Av^2 + Bv.$$

(E58)

To calculate the Reynolds number, we can rearrange the formula to find

$$\frac{mg}{v} = Av + B$$

(E59)

and use the following values in the two-parameter script from the appendix:

$$x = v,$$

(E60)

$$y = \frac{mg}{v},$$

(E61)

and

$$\sigma_y = \frac{mg}{v^2} \sigma_v.$$

(E62)

Finally, the Reynolds number can be calculated using

$$R = \frac{A_{best}}{B_{best}} v,$$

(E63)

where $v$ is the average velocity from all of your measurements.

### E.6 Solutions for Chapter 7

### 7.1 The mean deviation

The mean deviation is

$$\langle x - \langle x \rangle \rangle = \int_{-\infty}^{+\infty} x\rho \, dx - \int_{-\infty}^{+\infty} \langle x \rangle \rho \, dx.$$

(E64)

Using the definition of the mean and the normalization condition, we find

$$\langle x - \langle x \rangle \rangle = \langle x \rangle - \langle x \rangle \int_{-\infty}^{+\infty} \rho \, dx$$

$$= \langle x \rangle - \langle x \rangle \cdot 1 \tag{E65}$$

$$= 0.$$

## 7.2 The variance identity

The variance is

$$\langle (x - \langle x \rangle)^2 \rangle = \int_{-\infty}^{+\infty} (x - \langle x \rangle)^2 \rho(x) \, dx. \tag{E66}$$

Expanding the square, we find

$$\langle (x - \langle x \rangle)^2 \rangle = \int_{-\infty}^{+\infty} (x^2 - 2x\langle x \rangle - \langle x \rangle^2) \rho(x) \, dx$$

$$= \int_{-\infty}^{+\infty} x^2 \rho(x) \, dx - \int_{-\infty}^{+\infty} 2x\langle x \rangle \rho(x) \, dx + \int_{-\infty}^{+\infty} \langle x \rangle^2 \rho(x) \, dx$$

$$= \langle x^2 \rangle - 2\langle x \rangle \int_{-\infty}^{+\infty} x\rho(x) \, dx + \langle x \rangle^2 \int_{-\infty}^{+\infty} \rho(x) \, dx \tag{E67}$$

$$= \langle x^2 \rangle - 2\langle x \rangle^2 + \langle x \rangle^2$$

$$= \langle x^2 \rangle - \langle x \rangle^2.$$

## 7.3 The C–D Plane $\chi^2$

The two likelihoods are equal if the arguments of their exponentials are equal. For the *C–D* plane, we have

$$-\frac{(C - C^*)^2}{2\sigma_C^2} - \frac{(D - D^*)^2}{2\sigma_D^2} - \frac{\chi^2_{min}}{2}. \tag{E68}$$

Plugging in the following definitions:

$$C = \frac{A}{\sigma_A} + \frac{B}{\sigma_B},$$

$$C^* = \frac{A^*}{\sigma_A} + \frac{B^*}{\sigma_B},$$

$$\sigma_C = \sqrt{\frac{2}{1+\rho}},$$

$$D = \frac{A}{\sigma_A} - \frac{B}{\sigma_B}, \tag{E69}$$

$$D^* = \frac{A^*}{\sigma_A} - \frac{B^*}{\sigma_B}, \text{ and}$$

$$\sigma_D = \sqrt{\frac{2}{1-\rho}}$$

and neglecting the common $\chi^2_{\min}/2$ gives

$$-\frac{(C-C^*)^2}{2\sigma_C^2} - \frac{(D-D^*)^2}{2\sigma_D^2}$$

$$= -\left(\frac{1+\rho}{2}\right)\left(\frac{A-A^*}{\sigma_A} + \frac{B-B^*}{\sigma_B}\right)^2 - \left(\frac{1-\rho}{2}\right)\left(\frac{A-A^*}{\sigma_A} - \frac{B-B^*}{\sigma_B}\right)^2$$

$$= -\left(\frac{1+\rho}{4}\right)\left(\frac{(A-A^*)^2}{\sigma_A^2} + \frac{(B-B^*)^2}{\sigma_B^2} + \frac{2(A-A^*)(B-B^*)}{\sigma_A\sigma_B}\right) \tag{E70}$$

$$-\left(\frac{1-\rho}{4}\right)\left(\frac{(A-A^*)^2}{\sigma_A^2} - \frac{(B-B^*)^2}{\sigma_B^2} - \frac{2(A-A^*)(B-B^*)}{\sigma_A\sigma_B}\right)$$

$$= -\frac{1}{4}\left(\frac{2(A-A^*)^2}{\sigma_A^2} + \frac{2(B-B^*)^2}{\sigma_B^2}\right) - \frac{\rho}{4}\left(\frac{4(A-A^*)(B-B^*)}{\sigma_A\sigma_B}\right).$$

This simplifies to the desired form of

$$-\frac{(A-A^*)^2}{2\sigma_A^2} - \frac{(B-B^*)^2}{2\sigma_B^2} - \frac{\rho(A-A^*)(B-B^*)}{\sigma_A\sigma_B}. \tag{E71}$$

## 7.4   *The $\chi^2$ matrix*

We start by solving for the eigenvalues and eigenvectors of the matrix

$$M = \begin{bmatrix} 1 & \rho \\ \rho & 1 \end{bmatrix}. \tag{E72}$$

The eigenvalues are $(1 + \rho)$ and $(1 - \rho)$. The corresponding eigenvectors are

$$\frac{1}{\sqrt{2}}\begin{bmatrix} 1 \\ 1 \end{bmatrix} \text{ and } \frac{1}{\sqrt{2}}\begin{bmatrix} 1 \\ -1 \end{bmatrix}. \tag{E73}$$

To diagonalize $M$, we first create a matrix $R$ whose columns are the eigenvectors of $M$,

$$R = \frac{1}{\sqrt{2}}\begin{bmatrix} 1 & 1 \\ 1 & -1 \end{bmatrix}. \tag{E74}$$

We then use

$$D = R^{-1}MR, \tag{E75}$$

where $D$ is the corresponding diagonal matrix. In our case $R^{-1} = R$, so we have

$$D = \begin{bmatrix} (1+\rho) & 0 \\ 0 & (1-\rho) \end{bmatrix}. \tag{E76}$$

Next, we can rewrite our matrix form of $\chi^2$ as

$$\chi^2(A, B) = \chi^T M \chi = \chi^T RDR^{-1}\chi = \chi^T R(R^{-1}MR^{-1})R\chi, \tag{E77}$$

which is given by

$$
\begin{bmatrix} \chi_A & \chi_B \end{bmatrix} \cdot \frac{1}{\sqrt{2}} \begin{bmatrix} 1 & 1 \\ 1 & -1 \end{bmatrix} \cdot \begin{bmatrix} (1+\rho) & 0 \\ 0 & (1-\rho) \end{bmatrix} \cdot \frac{1}{\sqrt{2}} \begin{bmatrix} 1 & 1 \\ 1 & -1 \end{bmatrix} \cdot \begin{bmatrix} \chi_A \\ \chi_B \end{bmatrix},
\tag{E78}
$$

which can be rewritten as

$$
\begin{bmatrix} \sqrt{\frac{1+\rho}{2}}(\chi_A + \chi_B) & \sqrt{\frac{1-\rho}{2}}(\chi_A - \chi_B) \end{bmatrix} \cdot \begin{bmatrix} 1 & 0 \\ 0 & 1 \end{bmatrix} \cdot \begin{bmatrix} \sqrt{\frac{1+\rho}{2}}(\chi_A + \chi_B) \\ \sqrt{\frac{1-\rho}{2}}(\chi_A - \chi_B). \end{bmatrix},
\tag{E79}
$$

where we can finally define

$$
\chi_C = \sqrt{\frac{1+\rho}{2}}(\chi_A + \chi_B)
\tag{E80}
$$

and

$$
\chi_D = \sqrt{\frac{1-\rho}{2}}(\chi_A - \chi_B).
\tag{E81}
$$

## 7.5   The Poisson distribution

To find the average value of $n$, we sum over the distribution,

$$
\langle n \rangle = \sum_{n=1}^{\infty} n\rho(n)
$$

$$
= \sum_{n=1}^{\infty} n \frac{\mu^n e^{-\mu}}{n!}
$$

$$
= \sum_{n=1}^{\infty} \frac{\mu^n e^{-\mu}}{(n-1)!}
\tag{E82}
$$

$$
= \mu \sum_{n=1}^{\infty} \frac{\mu^{n-1} e^{-\mu}}{(n-1)!}
$$

$$
= \mu \sum_{n=0}^{\infty} \frac{\mu^j e^{-\mu}}{(j)!},
$$

where $j = n - 1$. This gives

$$
\langle n \rangle = \mu \sum_{n=0}^{\infty} \rho(j)
\tag{E83}
$$

$$
= \mu.
$$

To find the variance, we can use the variance identity from Problem 7.2,

$$
\langle (n - \langle n \rangle)^2 \rangle = \langle n^2 \rangle - \langle n \rangle^2.
\tag{E84}
$$

We already know that $\langle n \rangle^2 = \mu^2$, so all that is left is to find $\langle n^2 \rangle$. Summing over the distribution gives

$$
\begin{aligned}
\langle n^2 \rangle &= \sum_{n=1}^{\infty} n^2 \rho(n) \\
&= \sum_{n=1}^{\infty} n^2 \frac{\mu^n e^{-\mu}}{n!}.
\end{aligned}
\tag{E85}
$$

The trick is to split up $n^2$ using $n^2 = n(n-1) + n$. This gives

$$
\langle n^2 \rangle = \sum_{n=1}^{\infty} n(n-1) \frac{\mu^n e^{-\mu}}{n!} + \sum_{n=1}^{\infty} n \frac{\mu^n e^{-\mu}}{n!}.
\tag{E86}
$$

The second term is the definition of $\langle n \rangle$. We can rewrite the first term as

$$
\mu^2 \sum_{n=2}^{\infty} \frac{\mu^{(n-2)} e^{-\mu}}{(n-2)!}.
\tag{E87}
$$

Using $j = n - 2$, we find

$$
\begin{aligned}
\mu^2 \sum_{j=0}^{\infty} \frac{\mu^j e^{-\mu}}{j!} \\
= \mu^2.
\end{aligned}
\tag{E88}
$$

Combining all of this gives

$$
\begin{aligned}
\langle (n - \langle n \rangle)^2 \rangle &= \langle n^2 \rangle - \langle n \rangle^2 \\
&= (\mu^2 + \mu) - (\mu^2) \\
&= \mu.
\end{aligned}
\tag{E89}
$$

# Glossary

**Best fit line** The line that is closest to the measured data points, or the line with parameters that minimize the cost function.

**Central limit theorem (CLT)** The theorem that the distribution of the sum of many variables is well approximated by a Gaussian distribution.

**Chi-squared** Given by the formula $\chi^2 = \sum_i (y_i - y_{\text{model}i})^2 / \sigma_i^2$, where $y_i$ are the measured values, $y_{\text{model}i}$ are the model predicted values, and $\sigma_i$ are the uncertainties on the measured values.

**Cost function** A function based on the difference between data values and model values. The model parameter values that minimize the cost function are the best fit parameters.

**Degrees of freedom** The number of measurements (or data points) minus the number of model parameters.

**Deviation** The difference between a measured value and the mean or expected value.

**Error** Jargon for uncertainty. Typically, this should refer to experimental or calculational mistakes.

**Event** The name for the values that populate a histogram. For example, one bin in a histogram may contain twenty events.

**Gaussian** Also referred to as a normal distribution or a bell curve, given by the formula $G(y) = e^{-(y-\mu)^2/2\sigma^2} / \sqrt{2\pi\sigma^2}$, where $\mu$ is the mean and $\sigma$ is the standard deviation.

**Goodness of fit** A test used to determine whether a fit is good or not. For chi-squared analysis, goodness of fit is determined by whether the minimum value of $\chi^2$ is roughly equal to the number of measurements or data points in an experiment.

**Histogram** A plot that shows the number of events that correspond to each outcome. The width of each bin shows the range of values associated with that bin, while the height shows how many events fall within that range of values.

**Least squares** A technique where the best model is found by minimizing the cost function $C = \sum_i (y_i - y_{\text{model}i})^2$.

**Likelihood** Mathematically the same as the probability or probability density of an outcome, but a function of the parameters instead of the variables. While probabilities and probability density functions are used to predict outcomes for a given model, likelihoods are used to assess models for a given set of data.

**Linear regression** A family of techniques used to estimate the line of best fit for a data set. Similar to least squares fitting.

**Maximum likelihood estimation** A technique where the best fit parameters of a model are estimated by finding the parameters that make the measured data most probable.

**p-value** The probability of finding a certain value or greater, for a given probability density function.

**Parameters** The values in mathematical expressions that are not variables.

**Poisson distribution** The probability distribution for variables that take on only integer values, and whose distribution is fully described using the mean value. The mathematical form is $p(n) = \mu^n e^{-\mu} / n!$.

**Probability**   The portion of the total number of outcomes associated with a particular outcome.

**Probability density function**   For a given random variable or set of random variables, the probability density function gives the probability per infinitesimal range of the variable associated with any particular value of the variable. The integral of a probability density function over a range of variable values yields the probability of an outcome that falls within that range.

**Residual**   The difference between a measured value and a predicted value.

**Root mean squared deviation (RMSD)**   Given by the formula $\sqrt{\sum_i (y_i - \langle y \rangle)^2}$, where $y_i$ are the measurements and $\langle y \rangle$ is the mean value of $y$. Characterizes the spread of values.

**Standard deviation**   Jargon for the root mean squared deviation.

**Standard error**   The spread in mean values for sets of repeated measurements of a given size. Given by $RMSD/\sqrt{N}$, where RMSD is the root mean square deviation and $N$ is the number of measurements in each set.

**Uncertainty**   Characterizes the spread of measurements of the same variable. Random uncertainties are small fluctuations that cannot be experimentally controlled. Systematic uncertainties are associated with measurement techniques.

**Variables**   The values in mathematical expressions that may vary. In experimental terms, an independent variable is something that can be measured and changed by the experimenter, while a dependent variable can be measured by the experimenter, but changed only indirectly as a consequence of its relationship with an independent variable.

**Variance**   The mean squared deviation of a variable. Also the RMSD or standard deviation, squared.

# Books for Additional Study

Berendsen, Herman J. C., *A Student's Guide to Data and Error Analysis*, Cambridge University Press, 2011. (Section 7.4)

Bevington, Philip R. and Robinson, D. Keith, *Data Reduction and Error Analysis for the Physical Sciences*, McGraw-Hill, 3rd edition, 2003. (Sections 4.4, 6.3)

Flannery, Brian P., Press, William H., Teukolsky, Saul A., and Vetterling, William T., *Numerical Recipes: The Art of Scientific Computing*, Cambridge University Press, 3rd edition, 2007. (Sections 15.0–15.4)

Hughes, Ifan and Hase, Thomas, *Measurements and their Uncertainties: A Practical Guide to Modern Error Analysis*, Oxford University Press, 2010. (Chapters 6 and 8)

James, Frederick, *Statistical Methods in Experimental Physics*, World Scientific Press, 2nd edition, 2006. (Section 8.4)

Robinson, Edward L., *Data Analysis for Scientists and Engineers*, Princeton University Press, 2016. (Sections 2.6, 5.2, and 5.5.3)

Roe, Byron P., *Probability and Statistics in Experimental Physics*, Springer, 2nd edition, 2001. (Sections 13.1, 13.2, 14.1, and 14.2)

Taylor, John R., *An Introduction to Error Analysis: The Study of Uncertainties in Physical Measurements*, University Science Books, 2nd edition, 1996. (Chapter 12)

# Index